CELL BIOLOGY RESEARCH PROGRESS

A CLOSER LOOK AT FIBROBLASTS

CELL BIOLOGY RESEARCH PROGRESS

Additional books and e-books in this series can be found on Nova's website under the Series tab.

CELL BIOLOGY RESEARCH PROGRESS

A CLOSER LOOK AT FIBROBLASTS

JUSTIN O'SHANE
EDITOR

Copyright © 2020 by Nova Science Publishers, Inc.

All rights reserved. No part of this book may be reproduced, stored in a retrieval system or transmitted in any form or by any means: electronic, electrostatic, magnetic, tape, mechanical photocopying, recording or otherwise without the written permission of the Publisher.

We have partnered with Copyright Clearance Center to make it easy for you to obtain permissions to reuse content from this publication. Simply navigate to this publication's page on Nova's website and locate the "Get Permission" button below the title description. This button is linked directly to the title's permission page on copyright.com. Alternatively, you can visit copyright.com and search by title, ISBN, or ISSN.

For further questions about using the service on copyright.com, please contact:
Copyright Clearance Center
Phone: +1-(978) 750-8400   Fax: +1-(978) 750-4470   E-mail: info@copyright.com.

NOTICE TO THE READER

The Publisher has taken reasonable care in the preparation of this book, but makes no expressed or implied warranty of any kind and assumes no responsibility for any errors or omissions. No liability is assumed for incidental or consequential damages in connection with or arising out of information contained in this book. The Publisher shall not be liable for any special, consequential, or exemplary damages resulting, in whole or in part, from the readers' use of, or reliance upon, this material. Any parts of this book based on government reports are so indicated and copyright is claimed for those parts to the extent applicable to compilations of such works.

Independent verification should be sought for any data, advice or recommendations contained in this book. In addition, no responsibility is assumed by the Publisher for any injury and/or damage to persons or property arising from any methods, products, instructions, ideas or otherwise contained in this publication.

This publication is designed to provide accurate and authoritative information with regard to the subject matter covered herein. It is sold with the clear understanding that the Publisher is not engaged in rendering legal or any other professional services. If legal or any other expert assistance is required, the services of a competent person should be sought. FROM A DECLARATION OF PARTICIPANTS JOINTLY ADOPTED BY A COMMITTEE OF THE AMERICAN BAR ASSOCIATION AND A COMMITTEE OF PUBLISHERS.

Additional color graphics may be available in the e-book version of this book.

Library of Congress Cataloging-in-Publication Data

ISBN: 978-1-53616-977-5

Published by Nova Science Publishers, Inc. † New York

Contents

Preface		vii
Chapter 1	Molecular Mechanism of Photobiomodulation Associated with Fibroblasts *Sandra M. Ayuk and Heidi Abrahamse*	1
Chapter 2	The Role of Fibroblasts in Ovarian Cancer *Rosekeila Simões Nomelini,* *Isa Beatriz Carminatti Batista,* *Simone Paula Queiroz, Ana Carolinne da Silva* *and Eddie Fernando Candido Murta*	51
Chapter 3	Analysis of the Effect of an Andiroba, Copaíba and Guaraná Combination on *In Vitro* and *In Vivo* Scar Models *Euler Esteves Ribeiro Filho,* *Bárbara Osmarin Turra, Bruna Chitolina,* *Beatriz Sadigursky Cunha, Cibele Ferreira Teixeira,* *Verônica Farina Azzolin,* *Ednea Aguiar Maia Ribeiro, Euler Esteves Ribeiro,* *Raquel de Souza Praia, Juscimar Carneiro Nunes,* *Ivana Beatrice Mânica da Cruz* *and Fernanda Barbisan*	69

Chapter 4	Laser Therapy Associated with Guaraná as a Therapeutic Alternative on the Skin of Oxy-Inflammatory Metabolism *Daíse Raquel Maldaner, Cibele Ferreira Teixeira,* *Marta Maria Medeiros Frescura Duarte,* *Verônica Farina Azzolin,* *Ivana Beatrice Mânica da Cruz,* *Ednea Aguiar Ribeiro, Neida Luiza Pellenz* *and Fernanda Barbisan*	**109**
Index		**135**

Preface

A Closer Look at Fibroblasts first summarizes the molecular mechanisms of fibroblasts induced by photobiomodulation, reviewing current therapeutic approaches. Photobiomodulation, previously called low intensity laser irradiation, is a safe and efficient mechanism used to stimulate a positive response through absorbed light or light emitting diodes, and to reduce pain and inflammation promoting healing of the wound area.

Following this, the authors demonstrate the importance of fibroblasts in the process of development and progression of ovarian cancer, helping to establish new therapeutic management targeted towards cancer-associated fibroblasts.

In the penultimate study, the effects of the combination of andiroba, copaíba and guarana in the form of biphasic oil and emulsion is examined in the context of wound healing.

The concluding study focuses on the implications of the previously mentioned guarana in the context of skin aging and the oxy-inflammatory metabolism.

Chapter 1 - Photobiomodulation (PBM) previously called low intensity laser irradiation (LILI) is a safe and efficient mechanism used to stimulate a positive response through absorbed light (coherent light) or light emitting diodes (LED-non-coherent light), and to reduce pain and inflammation promoting healing of the wound area. PBM has been in existence for over

50 years; however, it is still not widely accepted due to the lack of understanding of the cellular and molecular mechanisms in various cells including fibroblasts. The mechanism of action of PBM with fibroblasts assumes that light is absorbed by mitochondrial chromophores to initiate cellular responses. The photons released dissociate inhibitory nitric oxide to increase electron transport, adenosine triphosphate (ATP) production and mitochondrial membrane potential (MMP). Furthermore, ion channels are activated and allows calcium to enter the cells and stimulate various signaling pathways leading to the activation of transcription factors which affect the protein expression, anti-inflammatory signaling, redox reactions, cell death mechanism, as well as stimulate processes to increase cell proliferation of various cell types, macrophage phagocytic activity, locomotion and maturation of fibroblasts. PBM decreases the inflammatory infiltrate, stimulating fibroblast proliferation and angiogenesis, and therefore recommended for wound healing processes. This chapter will summarise the molecular mechanisms of fibroblasts induced by PBM, reviewing current therapeutic approaches.

Chapter 2 - Ovarian cancer corresponds to a heterogeneous group of malignancies, and the majority is of epithelial origin. Ovarian neoplasms are divided into several histological subtypes. A classification that takes into account the ovarian carcinogenesis model divides ovarian tumors into two groups: type I and type II neoplasias. Type I neoplasms include serous, clear cell, endometrioid, mucinous, cancers, and have milder growth characteristics with slow growth, and most often grow from an identifiable precursor. On the other hand, type II neoplasias are characterized by neoplasias that develop rapidly, are more aggressive and classified as high grade, emphasizing that high grade serous ovarian carcinoma is the most common type II neoplasia, characterizing almost 75% of all cancers of epithelial origin.

Fibroblasts are responsible for providing structural integrity to most tissues, producing the tissue's own basement membrane, providing a protective barrier around the epithelium, thus contributing to the polarity, functionality and specificity of the epithelium. Such factors indicate that fibroblasts are critical parts during wound healing and inflammation processes.

Several recent studies have demonstrated the importance of the tumor microenvironment in the processes of evolution and progression of ovarian cancer, and stroma plays an important role in the mechanisms of cancer progression. The most common cell types in the tumor microenvironment are cancer-associated fibroblasts (CAFs), which are often present at all stages of the disease's evolution. The role of fibroblasts in the progression of ovarian cancer is complex, but some functions have already been well understood, such as the ability to produce extracellular matrix, chemokines, cytokines, growth factors, and also stimulate angiogenic recruitment events of pericytes and endothelial cells. In this way, fibroblasts are decisive components in the evolution of cancer.

CAFs activation is directly related to inflammation and recruitment of inflammatory cells to the stroma. Two cancer-associated fibroblasts markers, alpha-smooth muscle actin and fibroblast activation protein alpha, may be found in the stromal microenvironment of benign and malignant ovarian epithelial neoplasms, and to relate their tissue expression with prognostic factors in ovarian cancer.

The objectives of this chapter are to demonstrate the importance of fibroblasts in the process of development and progression of ovarian cancer, helping to guide new studies that establish new therapeutic management targeted to the CAFs. In addition, the authors will address stromal fibroblasts as potential therapeutic targets and the effect of chemotherapy treatment on fibroblasts.

Chapter 3 - *Introduction:* The skin is the largest organ of the human body and has the functions of coating, protecting and interacting with the external environment. Thus, the healing of the skin is a fundamental element for human health. The effectiveness of the healing process can further be enhanced by the development of products originating from plants with traditional use. Thus, in this study the authors searched for a healing compound based on the oils of andiroba (*Carapa guianensis*), copaíba (*Copaifera langsdorffii*) and guaraná (*Paullinia Cupana*), three plants of Amazonian origin that have potential healing effect through modulation of the inflammatory phase and proliferation of fibroblasts, with a consequent pro-cicatricial potential. Objective: To evaluate the effect of the combination of andiroba, copaíba and guarana (ACG®), in the form of biphasic oil and emulsion, through the analysis of *in vitro* wound healing models (using dermal fibroblasts) and *in vivo* (with earthworms *Eisenia fetida* as a model of tissue regeneration via surgical removal of the caudal region). *Methodology:* The antioxidant and genotoxic/genoprotective capacity of ACG® were evaluated in non-cellular tests. The line of dermal fibroblasts HFF-1 was cultured under standard conditions, and the stratch assay (*in vitro* model of skin lesion) was performed in the cells, that then were treated with ACG® in concentration of 2μL/mL and after 24/72 hours of incubation, and the cell migration and the proliferative, oxidative and inflammatory markers were analysed. In the *in vivo* model with Californian earthworms (*Eisenia fetida*), the animals had a caudal removal and were immediately treated with ACG®, in the form of biphasic oil and emulsion, and after 7 days the analysis of regeneration patterns was performed. Statistical analysis was performed by one-way analysis of variance followed by Tukey's post hoc. Tests with $p < 0.05$ were considered significant. *Results:* ACG® showed strong antioxidant activity, and was not genotoxic. In front of the cellular model, the product presented proliferative, antioxidant and anti-inflammatory action. In the *in vivo* tests, the biphasic oil was effective in the renegeration process. *Conclusion:* Despite the methodological limitations, the authors consider the data obtained to be relevant, once it was tested in the form of ACG® for the first time the combination of andiroba, copaíba and guaraná in cicatricial processes. The results show antioxidant, anti-inflammatory and pro-healing effects of the compound developed. Further studies must be performed to prove these results; however, the authors believe that in the future the development of a commercial product with clinical purposes may be a reality.

The results that will be presented come from a master's dissertation, presented at the Universidade Federal do Amazonas.

Chapter 4 - Skin aging is induced by genetic factors (chronological aging or intrinsic aging) and environmental or external factors, especially represented by sun exposure (photoaging, extrinsic or actinic aging). It is known that the cellular and molecular mechanisms of these factors are the same, that is, a superposition of the biological effects of extrinsic stressors occurs on intrinsic aging.

Therefore, it is emphasized that the organism as a whole undergoes a natural chronological aging process, simultaneously affecting all organs and tissues, because the cells possess a finite multiplication capacity, leading to cell senescence.

In: A Closer Look at Fibroblasts
Editor: Justin O'Shane

Chapter 1

MOLECULAR MECHANISM OF PHOTOBIOMODULATION ASSOCIATED WITH FIBROBLASTS

Sandra M. Ayuk and Heidi Abrahamse*, PhD
Laser Research Centre, University of Johannesburg,
Johannesburg, Gauteng, South Africa

ABSTRACT

(PBM) previously called low intensity laser irradiation (LILI) is a safe and efficient mechanism used to stimulate a positive response through absorbed light (coherent light) or light emitting diodes (LED-non-coherent light), and to reduce pain and inflammation promoting healing of the wound area. PBM has been in existence for over 50 years; however, it is still not widely accepted due to the lack of understanding of the cellular and molecular mechanisms in various cells including fibroblasts. The mechanism of action of PBM with fibroblasts assumes that light is absorbed by mitochondrial chromophores to initiate cellular responses. The photons released dissociate inhibitory nitric oxide to increase electron transport, adenosine triphosphate (ATP) production and mitochondrial

---

* Corresponding Author Email:habrahamse@uj.ac.za.

membrane potential (MMP). Furthermore, ion channels are activated and allows calcium to enter the cells and stimulate various signaling pathways leading to the activation of transcription factors which affect the protein expression, anti-inflammatory signaling, redox reactions, cell death mechanism, as well as stimulate processes to increase cell proliferation of various cell types, macrophage phagocytic activity, locomotion and maturation of fibroblasts. PBM decreases the inflammatory infiltrate, stimulating fibroblast proliferation and angiogenesis, and therefore recommended for wound healing processes. This chapter will summarise the molecular mechanisms of fibroblasts induced by PBM, reviewing current therapeutic approaches.

Keywords: fibroblast, lasers, photobiomodulation, woundhealing

1. Introduction

Photobiomodulation, (PBM), previously called low intensity laser irradiation is a safe, efficient, non-thermal and non-invasive application for acute and chronic pain treatment; in soft tissues, different inflammatory/neurological conditions as well as tissue repair, through absorbed monochromic or quasimonochromic light (coherent light) or light emitting diodes (LED-non-coherent light) (Tuner and Hode 2002). It has been in existence for over 50 years, but it is still not widely accepted due to the lack of understanding of the cellular and molecular mechanisms (Mester et al. 1967). Several research studies have been performed to give a better understanding of PBM therapy (Michael R Hamblin 2018; Tiina Karu 2010a; Khan and Arany 2015; Rojas and Gonzalez-Lima 2013). The wide use and effectiveness of PBM have converted many people to believe in the effectiveness of PBM treatment.

The description of the mechanism of action occurs both directly and indirectly. The direct stimulation promotes cell survival. In this case, chromophores such as cytochrome C oxidase (CCO) are absorbed by light, which releases photons that dissociate inhibitory nitric oxide to promote an increase in the mitochondrial function, generate transfer of electrons in the respiratory chain, mitochondrial membrane potential (MMP), adenosine triphosphate (ATP) as well as reactive oxygen species (ROS) production. These processes lead to the activation of transcription factors to initiate gene expression. Furthermore, ion channels are activated allowing calcium to enter the cells. In the cells, various signaling pathways are stimulated, leading to the activation of transcription factors which affect the protein expression, anti-inflammatory signaling, redox reactions, cell death mechanisms, as well as stimulate processes to increase cell proliferation of various cell types, macrophage phagocytic activity, locomotion and maturation of fibroblasts (Freitas and Hamblin 2016; Karu 2010a). For the indirect mechanism, intermediate systems modulate action in different areas of the body. The immune system or stem cells could be stimulated to provide relief in the affected areas of the body while decreasing pro-inflammatory cytokines as well as increasing anti-inflammatory cytokines (Johnstone et al. 2014).

PBM presents diverse biological responses depending on the laser parameters (Mester et al.1967; Rezende et al. 2007). Treatment is performed using specific wavelengths between 500-1100 nm and power output of 10-200 mW (Silveira et al. 2011). In cultured fibroblasts, PBM presents different effects in the morphological structure of the rough endoplasmic reticulum (RER) and mitochondria (Marques et al. 2004). Many *in vivo* and *in vitro* studies have utilized fibroblast cells to study the effect of PBM at different wavelengths and fluences (Table 1). Studies have shown that the application of PBM in the visible and near infrared (NIR) modulates wound healing through various cellular and biological mechanisms (Dawood and Salman 2013; Evans and Abrahamse 2009; Fernandes et al. 2018; Houreld et al. 2012; Huang et al. 2009; Navarro-Requena et al. 2018; Silveira et al. 2011; Skopin and Molitor 2009; Vitor et al. 2018; Zungu, Hawkins Evans, and Abrahamse 2009). This chapter will summarize the molecular mechanisms of fibroblasts induced by PBM, reviewing current therapeutic approaches.

Table 1. The effect of PBM at different wavelengths and fluences

	Wavelength (nm)	Doses/Fluence (J/cm^2)	Post irradiation time	Power output mW/cm^2	Spot size cm^2	Treatment duration	Effect of PBM on Fibroblast
(Zungu et al. 2009)	He-Ne 632.8 nm	5 or 16	1 or 24 h	3	9.1	45 mni 18 s and 2h 24 min 57s	Increased Ca^{2+}, MMP, ATP, cAMP at 5 J/cm^2
(Houreld et al. 2012)	Diode laser 660 nm	5 or 15	Immediately	100 mW/ 11mW/cm^2	9.1	7 min, 35 sec (5 J/cm^2) 22 min, 44 sec (15 J/cm^2)	Promotes mitochondrial enzyme activity increasing ATP and CCO
Skopin and Molitor (2009)	980 nm	3.1 -14.4/11.7	24-48 h	26-73/97	12.5 mm	2 min	Increased cell growth at low doses High doses –negative effect
(Vitor et al. 2018)	InGaAlP— 660 nm	2.5-6.2	6,12,24 h	5-15	-	10/20	Increase Col
(Fernandes et al. 2018)	InGaAlP— 660 nm	2.5 multiple application 5 Single 7.5 single application	24, 48, 72	Checked but not indicated	-	10-30 s	Increase proliferation and viability

	Wavelength (nm)	Doses/Fluence (J/cm^2)	Post irradiation time	Power output mW/cm^2	Spot size cm^2	Treatment duration	Effect of PBM on Fibroblast
(Chen et al. 2009)	810 nm	3	-	-	-	-	Activates nuclear factor kappa-light-chain-enhancer of activated B cells (NF-kB)
(George et al. 2018)	636 and 825 nm	5, 10, 15, 20, 25 J/cm^2		74 and 104 mW 8.26 and 11.45 mW/cm^2	9.08	605.28 - 3026.4 s 436.5 - 2182.5 s	Low ROS generation at 636 at all fluences High ROS at 825 except 10 which was low Optimal ATP production until 15 J. ATP 825>636
(Moore et al. 2005)	665 and 675 nm						Increase cell proliferation at 665, 675 while 810 showed an inhibitory response
(Hawkins and Abrahamse 2006)	632.8 nm	0.5, 2.5, 5, 10, and 16 J/cm^2	48 h	3 mW/cm^2		121 s to 88 min	Increased mitochondrial activity at 5 J
(Chen et al. 2011)	810 nm	0.003, 0.03, 0.3, 3 and 30 J/cm^2	6-10 h	1 mW/cm^2 to 30 mW/cm^2		7 s-5 min	Increase mitochondrial respiration activates, increase redox-sensitive to NFkB, ROS and ATP

Table 1. (Continued)

	Wavelength (nm)	Doses/Fluence (J/cm^2)	Post irradiation time	Power output mW/cm^2	Spot size cm^2	Treatment duration	Effect of PBM on Fibroblast
(Alexandratou et al. 2002)	647 nm	1.5 mJ/cm2		0.1 mW		15 s	Increase Ca2+ production, ROS and MMP
(Gkogkos et al. 2015)	Nd:YAG laser 1064 nm	2.6 J/cm^2, 5.3 J/cm^2, 7.9 J/cm^2, and 15.8 J/cm^2	24, 48, and 72	0.5 W		20, 40, 60, or 120 s	Increase proliferation and EGF
(Chen et al. 2000)	Nd:YAG laser	50–150 mJ		1.0–3.0 W		10 s	Degeneratively cytomorphologic change to cell death
(Pourzarandian et al. 2005)	Er:YAG laser	1.68 -5.0 J/cm^2.	-	-	-	-	Stimulates wound healing at 3.37 J/cm^2
(Poon, Huang, and Burd 2005)	Q-switched frequency doubled Nd:YAG 532 nm	0.8 J/cm^2 0.3, 0.5, 0.8, 1.0, 1.5, 2.0, 3.0, and 4.0 J/cm^2	24 h	-	2, 3, 4, 5 and 6 mm	-	Increased SCF, HGF and b-FGF gene expressions
(Yeh et al. 2017)	660 nm	8 J/cm^2	-	70 mW/ 15.17 mW/cm^2	-	-	Inhibit arecoline-mediated effects via the cAMP
(Taflinski et al. 2014)	LED 420 nm	-	24	50 mW/cm^2	-	-	Non-toxic doses inhibits TGF-β1-

	Wavelength (nm)	Doses/Fluence (J/cm^2)	Post irradiation time	Power output mW/cm^2	Spot size cm^2	Treatment duration	Effect of PBM on Fibroblast
(Hakki and Bozkurt 2012)	940 nm	20 ms-1 ms	-	0.3-2 W	-	-	Induction of growth factors mRNA expressions
(Shingyochi et al. 2017)	CO_2 laser system 10600 nm	0.1 J/cm^2 (52.08 mW/cm^2, 2 sec), 0.5 J/cm^2 (52.08 mW/cm^2, 10 sec), 1.0 J/cm^2 (52.08 mW/cm^2, 20 sec), 2.0 J/cm^2 (52.08 mW/cm^2, 40 sec), and 5.0 J/cm^2 (520.83 mW/cm^2, 10 sec).	-		35 mm	-	Increased proliferation and migration.- Regulated kinase (ERK), and Jun N-terminal kinase (JNK).
	Wavelength (nm)	Doses/Fluence (J/cm^2)	Post irradiation time	Power output mW/cm^2	Spot size cm^2	Treatment duration	Effect of PBM on Fibroblast
(Nowak et al. 2000)	Superpulsed CO_2 laser energy	2.4, 4.7, and 7.3 J/cm^2	-	-	-	-	Enhanced fibroblast and stimulated bFGF secretion and to inhibit TGF-β1 secretion
(Esmaeelinejad and Bayat 2013)	He-Ne 632.8	0.5, 1, and 2 J/cm^2	-	0.66	-	-	Stimulated the release of IL-6 and bFGF
(Ling et al. 2014)	632.8 nm	0 1 J/cm^2	24 h	10 mW, 12.74 mW	-	-	Increased FOXM1,

2. Fibroblasts and Wound Healing

Fibroblasts are located in the dermis of the skin. They are derived from fibrocytes, and commonly found in the connective tissues in animals (Alberts et al. 2002). They are characteristically spindled-shape comprising of a nucleus with one or two nucleolus, mitochondria, several microtubules and RER (Figure 1) (Phan 2008). Fibroblasts constitutes collagen, lattice and elastic fibers, glycosaminoglycan, glycoproteins in the exocrine matrix, and thymic stromal lymphopoietin cytokines. Fibroblast comprise of four human sub-populations: the dermal fibroblasts which originates from two families; the upper dermal which is involved with hair follicles and the lower which is involved with adipocytes, repair as well as extracellular matrix (ECM) synthesis (Driskell et al. 2013). Fibroblasts play a major role in the production and maintenance of the ECM. They organize and regulate other cell types. In addition, they form an integral part in homeostasis, epidermal morphogenesis, as well as differentiation. Fibroblasts have the ability to differentiate into myofibroblasts. Myofibroblasts express α-smooth muscle actin (α-SMA) and are characteristic of both smooth muscles and fibroblasts, which enables the cells to contract and ensure wound closure (Plikus et al. 2017).

Growth factors stimulate fibroblasts activity and differentiation. Some of these include fibroblasts growth factor (FGF), vascular endothelial growth factor (VEGF), platelet-derived growth factor (PDGF), and transforming growth factor-β1 (TGF-β). TGF-β1 mediates uncontrolled proliferation of myofibroblasts leading to increased expression of α-SMA. TGF-β1 increases contraction of myofibroblasts by inducing the expression of collagen III, α-SMA and fibronectin (Krassovka et al. 2019). In chronic non-healing wounds, fibroblasts present decreased proliferation, irregular pattern of the release of cytokines, early senescence, abnormal tissue inhibitor of metalloproteinases (TIMP) and Matrix metallopeptidases activity, etc. (Cook et al. 2000; Wall et al. 2008). Through paracrine signaling, the structure of keloid fibroblasts might be altered (Ashcroft et al. 2013).

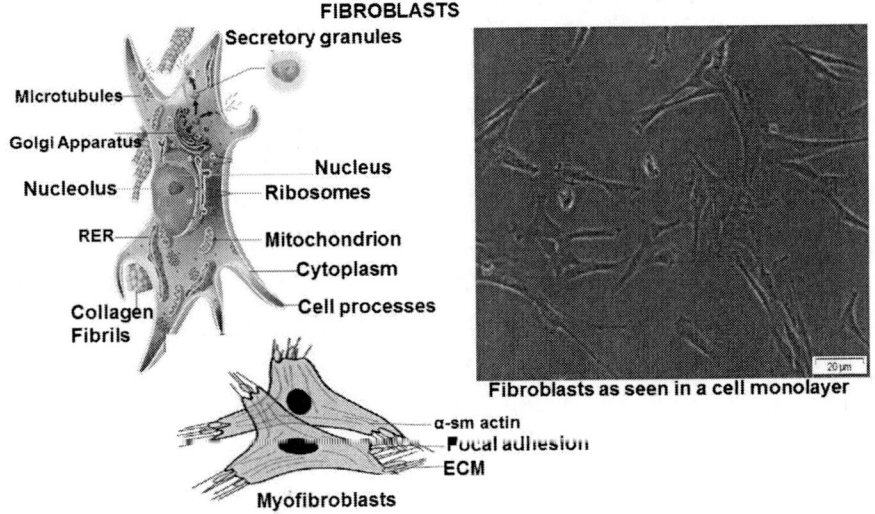

Figure 1. Structure of Fibroblast. Schematic diagram showing the various components of a fibroblast cell, its differentiated cell (myofibroblast) and morphology of fibroblast in cell culture.

2.1. Wound Healing

A condition whereby the skin experiences trauma, abrasion, ulcers, or introduction of pierced objects that disrupt the normal tissue structure *in vivo* or produces a central scratch in an *in vitro* environment is termed a "wound" (Enoch and Leaper 2008). Wound healing following homeostasis is a multifaceted process summarized into three overlapping phases including inflammation, proliferation and remodeling. At different stages, there are interactions of various cellular components, cytokines, chemokines, clotting factors and growth factors (Barrientos et al. 2008; Camposet al. 2008; Gurtner et al. 2008; Li et al. 2007). These interactions enhance the cell migration and proliferation, tissue granulation and formation, collagen synthesis, angiogenesis and tissue remodeling (Falanga 2005; Gosain and DiPietro 2004; Mathieu et al. 2006). Also involved in the wound healing process includes inflammatory mediators, neutrophils, platelets, monocytes, macrophages, endothelial cells, keratinocytes, endothelial cells and lymphocytes, as well as fibroblasts (Chrysanthopoulou et al. 2014). Fibroblasts function in wound contraction, breakdown of fibrin clot, deposition of ECM components, and remodeling of new ECM (Bainbridge 2013; Gosain and DiPietro 2004; B. Hinz 2016). For effective healing to occur, the process must occur in the correct order and time (Mathieu et al. 2006). Factors that can disrupt or delay the normal healing process include diabetes, hypoxia, collagen disorders, autoimmune disorders, malnutrition, uremia, malignancies, jaundice, etc. (Guo and DiPietro 2010; Rasik and Shukla 2000). Regulation of these factors through the collaboration with different cells, growth factors, cytokines, as well as the components of the ECM promote effective healing (Barrientos et al. 2008; Penn et al. 2012; Rolfe and Grobbelaar 2010; Werner and Grose 2003; Wong and Crawford 2013).

The phases of wound healing are coordinated in a way that there is proper communication between the inflammatory mediators and wounded cells through movement of blood vessels, fibroplasia and re-epithelialization. The inflammatory phase of wound healing takes 24-48 h. Different cells including platelets, polymorphonuclear leukocytes (PMNs), macrophages, monocytes, fibroblasts, keratinocytes as well as cytokines and growth factors are stimulated to facilitate wound healing and tissue regeneration (Beldon 2010; Enoch and Leaper 2008; George et al. 2006; Gurtner et al. 2008). In the inflammatory phase of wound healing, cytokines and chemokines initiate fibroblasts migration to the wound matrix (El Ghalbzouri et al. 2004). Platelets release cytokines to dilate the blood vessels, and improve permeability into the capillaries, to ease the migration of PMNs and growth factors. Platelets are also involved in the coagulation process. Lack of macrophages at the final phase of inflammation, disrupts the wound healing process, retarding fibroblast proliferation and giving way to the proliferation phase. It also leads to improper angiogenesis, poor fibrosis and poor debris removal (Enoch and Leaper 2008; Newton et al. 2004). The process might take longer in chronic wounds (Willenborg et al. 2010).

In the proliferative phase, the fibroblasts migrate to the wound. It involves overlapping events including angiogenesis, formation of tissue granules, fibroplasia, epithelialization, contraction as well as collagen deposition (Guo and DiPietro 2010). Proliferation is supported through the formation of granular tissue, ECM contraction, chemicals such as nitric oxide (NO), the lack of contact inhibition as well as deposited fibronectin which has been cross-linked with fibrin during the inflammatory phase (Falanga 2005; Lorenz and Longaker 2008; Witte and Barbul 2002). Components of the ECM are found in all the three phases of wound healing and are involved in tissue remodeling, angiogenesis and rapid scaffold breakdown (Badylak 2002). They provide biochemical and mechanical support creating a distinct microenvironment for the cells and prompts signal transmission. This is very important for surrounding cells including fibroblasts and chondrocytes. Fibroblasts upset the arrangement of collagen fibers and assist in forming ligaments, tendons and connective tissues that bind various organs (Halayko et al. 2008; Rauch 2004). The movement of keratinocytes, cell proliferation and differentiation would promote re-epithelialization, critical in wound repair (Santoro and Gaudino 2005). Fibroblasts release the components of the ECM such as glycosaminoglycan (GAG) and proteoglycans (PG) as well as collagen with specific signals from keratinocytes. Keratinocytes release enzymes, growth factors and cytokines to the wound site to facilitate cell proliferation (Terblanche et al. 2009; Tomic-Canic 2005). Fibroblast initiate collagen synthesis and reorganization of provisional ECM (Beldon 2010; Enoch and Leaper 2008). Wound contraction begins 8-10 days as the fibroblasts differentiate into myofibroblasts. As the myofibroblasts move towards the wound edge, they become attracted to fibronectin and growth factors. Furthermore, they contract, reduce the size of the wound and stick to themselves with the assistance of desmosomes (Hinz 2016). Additionally, they are able to associate with other molecules across the ECM (Mirastschijski et al. 2004). Towards the final wound contraction, there is decreased movement and proliferation of fibroblasts (Beldon 2010).

The remodeling phase is important for maintaining the tissue structure at the final phase in tissue repair and scar formation increasing the wound strength. It occurs between 3-21 days. Fibroblasts and keratinocytes release proteolytic enzymes in the ECM leading to the reduction of fibronectin levels, breakdown collagen type-I (Col-I), and other matrix proteins (Alberts et al. 2002; Bogaczewicz et al. 2005; Herouy et al. 2000). The ECM undergoes degradation to ensure a successful tissue remodeling (Agren and Werthen 2007).

The role of fibroblasts in wound healing is critical (Rhee 2009). It utilizes GAG, elastin (El), fibronectin (FN) and laminin (LMN) to produce collagen, an important component that preserves the structural integrity of the cell around the ECM. During the repair process of an injury, fibroblasts migrate from the bone marrow to the blood. They stimulate several growth factors and secretions to assist in their function. These cellular and biological processes include cell migration, proliferation, differentiation as well as the formation of the ECM (Desmoulière et al. 2005; George et al. 2006). Recent studies have shown that fibroblast cell proliferation and migration are key determinants of tissue repair following an injury and that the coordination of tissue-scale is facilitated by the interdependence of ECM deposition and cell proliferation, creating novel therapeutic interventions to promote skin regeneration (Rognoni et al. 2018). PBM in the visible and NIR have been shown to stimulate this cells and other surrounding cells (Hamblin 2018), assisting in wound healing by keeping the connective tissue together as well as facilitating the production of collagen (Ayuk et al., 2010). This light interaction with endogenous photoreceptors, ion gate channels, chromophores such as CCO, opsins and nitrosated and flavo proteins stimulate photobiomodulation (Chung et al. 2012; Freitas and Hamblin 2016).

3. PHOTOBIOMODULATION

3.1. History of Low Intensity Laser Irradiation (LILI)

Einstein (1917) was the first to describe the acronym LASER (Light Amplification by Stimulated Emission of Radiation). Further studies involved stimulated emission and gas molecules where scientists succeeded to develop a pulse system called MASER (Microwave Amplification by Stimulated Emission of Radiation), which worked with ammonia only (Gordon, Zeiger, and Townes 1955). Furthermore, Schawlow and Townes (1958) thought of modifying into an optical and infrared (IR) MASER. However, these contributed efforts did not get the first laser working but emphasized on the principles which Maiman (1960) used to realize the first working laser. Maiman (1960) designed a pulse-operated ruby laser for the emission of light in the visible red spectrum. The shortcoming of this laser was that it had only three energy levels meanwhile a functional continuous system would need four energy levels. Finally, Ali Javan, William Bennet and Donald Herriot developed the first continuous gas, helium-neon (He-Ne) laser (Javan et al. 1961). In 1964 by Nikolaj Basov and Alexander Prokhorov established continuous output system, whom together with Charles Townes won a Nobel Prize. This innovation brought about the existence of several lasers now applicable in the military, industrially, commercially, scientifically, and medically (Javan et al.1961). Furthermore, Bohr (1921) demonstrated the laser principles based on the quantum theory. The principle stated that stimulated emission increases along the same axis when spontaneous emission decreases implying that longer wavelengths emits photons with lower energy.

Phototherapy existed around 1900s when Niel Fiensen in 1903 won a Nobel Prize for using light radiation to treat lupus vulgaris. However, the use of the word LASER raised several questions over its electromagnetic nature. This was because high-energy electromagnetic radiations produced gene toxicity and DNA damage. Nonetheless, high-powered lasers in the visible and NIR spectrum, used in treatment are non-thermal and non-invasive. et al. (1967) for the first time used a ruby laser at 694 nm to prove that lasers at low doses could have a biostimulatory effect on mice. Upon irradiation, there was improved wound healing and hair growth (Mester et al. 1968; Tibbs 1997). He also found that phototherapy could be efficient in stimulating animals and used in clinical studies (Kov et al. 1974). Knowing the usefulness of this treatment especially in surgery, both coherent (laser) and non-coherent (LED) light sources have been implemented for treatment (Chaves et al. 2014).

3.2. Definition of PBM

Several appellations have been used in the past for the treatment with low doses of light such as biostimulation or photobiostimulation (PBM), low-level/intensity light or laser irradiation (LLLI/LILI) or therapy (LLLT/LILT), as well as cold/cool or soft laser. The usage of the term PBM for this treatment has gained approval worldwide. This application has been used to decrease inflammation, produce analgesia and improve tissue regeneration in osteoarthritis, tendinopathies, nerve injuries, and wound healing (Chung et al. 2012) and led to the definition of PBM. PBM is defined as a non-thermal/non-invasive process involving endogenous chromophores such as rhodopsin (rod cells in retina), seven dehydrocholesterol (vitamin D3 in skin), photopsin (cones cells in retina), CCO (mitochondria), flavins, porphyrins, NADPH, melanopsin (photosensitive ganglion cells in retina), etc., to produce photobiological responses (Tiina Karu 1989).

Apart from the conventional approach of using biological and pharmaceutical agents for wound treatment, PBM is a modality that has been beneficial to aid wound closure through cellular and molecular mechanisms. These include the promotion of epithelial cell migration, neutrophil and macrophage migration and function, synthesis of extracellular matrix and wound contraction, as well as fibroblast proliferation. These mechanisms occur through oxidation and reduction reactions releasing ROS (Arany 2012; Chen et al. 2011) or through structural changes of the molecule (Liebert et al. 2014). Free radicals produced are classified as hydrogen peroxide, superoxide and hydroxyl radicals. Electron transfer generates nitrogen oxide (NO) which produces positive response to PBM by reacting with superoxide (O^{-2}) released by inflammatory cells to form peroxynitrite ($ONOO^-$) (Assis et al. 2012). Different responses are produced based on the dose applied (biphasic dose response) but these still needs to be clearly understood (Huang et al. 2011).

3.3. Laser Parameters

The basis of biological responses of PBM depends on the laser parameters. Parameter such as light and dose are very crucial in PBM response. However, consideration in selecting variables is important or mandatory for chances of a successful treatment. Such variables include power output, wavelength, energy density, energy, total power, spot size, pulse structure, treatment repetition regime, as well as tissue absorption characteristics that contribute to a positive outcome. In addition, the use of combination of more than one wavelength, treatment duration, and type of laser application is a consideration to a lesser extent (Chung et al. 2012; Jenkins and Carroll 2011). To achieve a certain outcome the parameters are regulated over a range of values. Positive responses particularly have been demonstrated in the visible spectra region between 600–700 nm and in the near infrared region between 770-1200 nm. Certain grey areas have produced poor results due to the penetration depth. The longer the wavelength the shorter the penetration depth with less absorption and scattering (Wang and Jacques 1992). Light penetration is dependent on the scattering, absorption and structural property of the molecule. Calabrese and Mattson (2017) defines hormesis as a determining factor of the performance of many indicators in the biological process. These are characterized by the simultaneous stimulation of various independent cellular/biological processes. They are also regulated by signaling pathways, or multiple interacting receptors to produce a certain outcome. This concept has been used to determine the parameters of laser according to Arndt-Schulz law, which states that "the action of substances may vary according to whether the dose is high to kill, moderate to inhibit or low to stimulate" (Martius 1923). Arndt-Schulz law generally describes the dose dependent effect of PBM and has been used for many years to explain the interaction of cells and light.

The mechanism of application of PBM is based on the law implying that at low energy levels absorption of photons by CCO triggers the primary effects (Karu 2010b) such as the production of ATP and consumption of oxygen leading to the dissociation of NO, production of growth factors, activation of transcription factors, and secondary messenger pathways (Freitas and Hamblin 2016). The amount of energy released and rate plays a vital role in the PBM response (Huang et al. 2009; Huang et al. 2011). Mester et al. (1985) reviewed the biomedical effects of laser application. In their study, the researchers found that low-energy (J/cm^2) laser irradiation had a stimulatory effect on cells meanwhile high-energy laser presented an inhibitory response. Zein et al. (2018) conducted a systematic review and meta-analysis on cultured cells *in vitro* and on different tissues *in vivo* to evaluate the efficacy of light parameters in PBM. They studied different cells and tissues with respect to the number of mitochondria present and dose response. The study found that cells and tissues with high numbers of mitochondria responded to lower doses of light compared to those with lower numbers. In addition, the fluence administered in an *in vitro* study would determine the success or failure compared to the irradiance. The chapter will review the effect of the parameters used in fibroblast cells in response to PBM (Table 1).

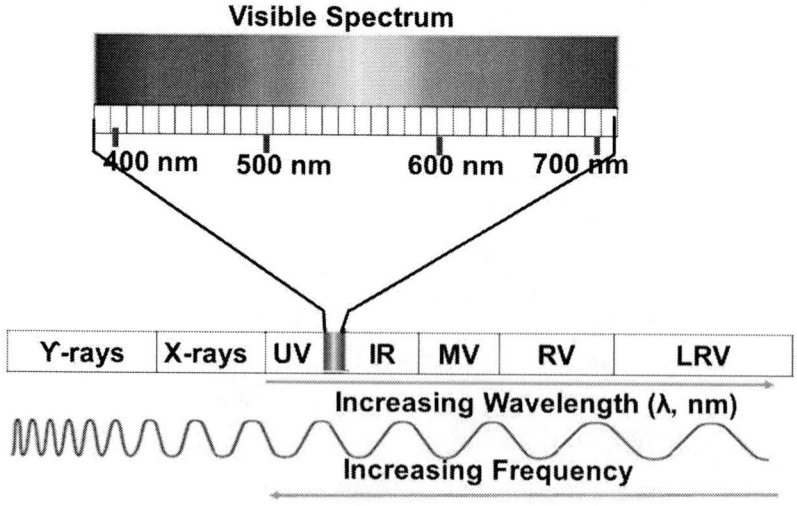

Figure 2. Wavelengths used in phototherapy as represented in the electromagnetic spectrum. Frequencies range from gamma-rays (Υ), X-rays, Ultraviolet (UV), radiowaves (RV) and long RV. The wavelengths from the visible light spectrum (400-700 nm) and infrared (IR, 800-1,100 nm) (Adapted from Foresman, 2009).

3.3.1. Power Density (Pd)

Power density (Pd) is the output power (P) which is emitted per unit area. It is the beam area or spot size (A) of the incident sample represented in watts or milliwatts per centimetre squared (W/cm^2 or mW/cm^2) (Huang et al. 2009). The calculation is as follows:

Pd = P (W)/A (cm^2)
(Where A is the area of the laser beam in centimetres squared)

3.3.2. Energy Density

The energy density denotes the amount of energy that is applied to the sample per unit area (A). It signifies the measure of the dose or fluence. To achieve this, you multiply the power density (W/cm^2) by the exposure time (t). The energy is represented in Joules (J) (Huang et al., 2009). The calculation is as follows:

Fluence or dose (J/cm^2) = Energy (J)/Area (A) =
Power density (W/cm^2) x time (s)

---

where J, represents the energy in joules and A the cross sectional beam area in cm^2.

4. MOLECULAR MECHANISMS

4.1. Mitochondrial Activities

4.1.1. Cytochromic Oxidase (CCO)

The inner mitochondrial membrane consist of 5 membrane protein complexes namely NADH dehydrogenase (Complex I), succinate dehydrogenase (Complex II), cytochrome c reductase (Complex III), CCO (Complex IV), ATP synthase (Complex V), and two freely diffusible molecules, ubiquinone and cytochrome c, which facilitates electron movement from one complex to the other. CCO is described as a large transmembrane multicomponent protein, comprising of a 13 proteins, three binuclear copper (CuA) and heme (a3-CuB) groups as well as zinc and magnesium located at the terminal electron transfer chain to promote electron transfer to oxygen (Karu and Kolyakov 2005; Srinivasan and Avadhani 2012). Its main function is oxidation and reduction.

CCO is a photoacceptor in photobiomodulation (Eells et al. 2003; Poyton and Ball 2011; Wong-Riley et al. 2005). Many studies have represented CCO as the most important chromophore for light absorption. CCO consists of two absorption bands in the visible spectrum at approximately 660 nm as well as 800 nm in the NIR spectrum (Chung et al. 2012). Over 50% of light absorption that occurs in the range of 800-850 nm to a lesser extent with oxy and deoxyhaemoglobin (HbO_2 and Hb) (Chance et al. 1997). There is a correlation between the phototherapeutic action and the absorption spectra of various CCO oxidation states (Karu and Afanas' eva 1995; Karu et al. 2008; Wang et al. 2016; Wong-Riley et al. 2005). CCO plays a very important role with diverse functions in PBM. According to research, red light or NIR enhanced cell proliferation in the mitochondrial chain, however, metabolic changes of PBM are not always associated with CCO (Lima et al. 2019). CCO also maintains the movement of protons across the inner mitochondria under normal physiological conditions. The photobiostimulation of CCO increases electron transfer which leads to increase ATP production through oxygen reduction, activating downstream processes such as DNA and RNA synthesis, proteins, enzymes etc. necessary for tissue regeneration (Yu et al. 1997). CCO also produces NO, which is important in intra and extra cellular pathways (Poyton and Ball 2011). PBM increases CCO/NO activity by dissociation of NO from a32+NO- complex of CCO (Hayashi et al. 2007; Karu and Kolyakov 2005). Houreld et al. Abrahamse (2012) found that PBM (660 nm, $5/15J/cm^2$) increased complex IV activity in stressed fibroblast cells *in vitro* which supports the hypothesis CCO is a photoacceptor in the visible light spectrum.

4.1.2. Adenosine Triphosphate (ATP)

All living things need to make Adenosine Triphosphate (ATP) or cellular energy. ATP is the source of energy for every cell in the body. During cellular respiration, an enzyme, cytochrome c oxidase (CCO), helps oxygen bind with nicotinamide adenine dinucleotide and hydrogen (NADH). This produces ATP synthase (necessary for ATP production). Red light can penetrate 8 to 10 millimeters into the skin, which is enough to affect the body on a cellular level. It then stimulates the production of ROS that helps to send signals from mitochondria to the nucleus. Intracellular ATP is released when there is an injury which affects the plasma membrane or cause apoptosis/necrosis. ATP production triggers movement of various cells to the injury site (Chen et al. 2006; Elliott et al. 2009; Janssen et al. 2009). Apart from the intracellular ATP from an injury, ATP could also be released via gap junction channels (Lohman et al.2012; Sáez et al. 2003). Red light also breaks the bond of nitric oxide with CCO, allowing it to bind with oxygen to make ATP synthase. ATP synthase converts the proton energy into ATP. ATP produces energy and serves as an intracellular signaling molecule, allowing cells and tissue communication throughout the body (Burnstock 2009; Pinheiro et al. 2013; Schwiebert and Zsembery 2003). Extracellular ATP from fibroblasts would facilitate hypoxia-induced adventitial fibroblast growth in various tissues as well as stimulates Ca^{2+} waves and gene expression (Gerasimovskaya et al. 2002; Janssen et al. 2009). PBM modulates mitochondrial response in fibroblast cells. Zungu et al. (2009) investigated the potential effect of laser irradiation on hypoxic and acidotic human skin fibroblasts and found that PBM (He-Ne, 632.5 nm, 5 J/cm^2) modulated an increase in intracellular Ca^{2+} which led to an increase in MMP, ATP and cAMP *in vitro*. They concluded that PBM is able to restore homeostasis in injured cells. Houreld et al. (2012) examined the enzyme activities of mitochondrial complexes I, II, III, and IV using diode laser (660 nm, 5/15 J/cm^2). They also found an increase in ATP immediately after irradiation in both normal and diabetic human skin fibroblast cells.

4.1.3. Cyclic Adenosine Monophosphate (cAMP)

cAMP is a signal transduction mediator of many neuropeptides and hormones. Some cAMPs modulate the ECM composition. On the cell surface of hormones, they transmit signals to initiate the activation of cAMP dependent protein kinase A (PKA). Transmission occurs through guanine nucleotide-binding (G)-protein-coupled (G) protein receptors. cAMP interferes with the functioning of TGF-β. The levels of intracellular cAMP activity are determined when there is a balance between adenylate cyclase and cyclic nucleotide phosphodiesterase activities (Montminy 1997). Cellular effects of cAMP occurs through the activation of three different kinds of sensors: cAMP-dependent protein kinase A (PKA) which phosphorylates and activates cAMP response element-binding protein (CREB). CREB binds to CRE domain on DNA, and in turn activates genes (Tasken and Aandahl 2004). Cyclic nucleotide-gated channels (CNGC) and exchange proteins are directly activated by cAMP (Epac) (Bos 2003; Zagotta and Siegelbaum 1996). Yeh et al. (2017) investigated the molecular mechanisms of PBM (660 nm, 8 J/cm^2) on the expression of arecoline mediated fibrotic marker genes in human gingival fibroblasts (HGFs). The authors demonstrated for the first time that, PBM through the cAMP signaling pathway suppresses changes in arecoline-associated fibrotic marker gene expression and inhibits the transcriptional activity of CCN2 in HGFs. They also found that the activation of cAMP expression inhibits arecoline-mediated fibrotic marker genes meaning that upregulating the cellular cAMP level blocks arecoline-induced fibrotic marker gene expression as well as the effect of PBM. This may have occurred through activation of the cAMP signaling pathway (Yeh et al. 2017).

Zungu et al. (2009) also reported that PBM (632.5 nm, 5 J/cm^2) enhances cAMP production in fibroblast cells. Schiller et al. (2010) evaluated the *in vitro* effects of increased intracellular cAMP levels on various key effector functions of fibroblasts. Their study presented vital roles of cAMP in TGF-β as well as TGF-β-mediated expression of several ECM components such as wound closure and contraction. There was an elevation of intracellular cAMP by artificial stimulators such as the fungal drug forskolin or Bt$_2$cAMP, which presented multidirectional functional consequences, either stimulatory or inhibitory, on Smad-dependent genes, collagen synthesis and contraction, production of HA, and fibroblast migration. Most importantly, the data provided profound evidence for a functional interaction between the cAMP/CREB and TGF-β signaling pathway in human dermal fibroblasts (HDF) (Schiller et al. 2010). It was concluded that increased cAMP levels modulate TGF-β /Smad-induced expression of ECM components and other key fibroblast effector functions. Furthermore, the cAMP pathway was identified as a potent but differential and promoter-specific regulator of TGF-β-mediated effects involved in ECM homeostasis.

4.1.4. Reactive Oxygen Species (ROS)

ROS activates the mitogen-activated protein kinase (MAPK) family, namely extracellular signal-regulated kinase (ERK), c-Jun NH2- terminal kinase (JNK) and p38. This initiates activator protein 1 (AP-1), a transcription factor which plays a major role in the transcriptional regulation (Quan et al. 2013). MMP when it is above or below normal values generate ROS. Not much is known whether the ROS generated are identical. George et al. (2018) investigated the effect of red light (636 nm) and near infrared laser (825 nm) using 5, 10, 15, 20, 25 J/cm^2 on the generation of ROS in primary dermal fibroblasts. They found that low amount of ROS was generated at 636 nm in all the fluences. On the otherhand, the amount of ROS produced at 825 nm was higher except for 10 J/cm^2, which was low. These authors suggested that the cellular redox status responds quite differently to certain fluences of 825 nm with some unknown mechanisms. The outcome from the study stated that ROS production within biological systems are more dependent on the wavelength of the laser rather than its fluence (George et al. 2018). There is small amount of ROS generated from PBM action on CCO in normal cells resulting from the increase in MMP (Chen et al. 2009). In embryonic fibroblasts, PBM (810 nm) released a small amount of ROS, which activated NF-kB, a redox activated transcription factor and also increased ATP (Chen et al. 2009; Chen et al. 2011). Using a single fibroblast cell, Alexandratou et al. (2002) observed in a confocal microscope that PBM at 647 nm increased ROS generation.

4.1.5. Calcium Ions (Ca^{2+})

Increased ATP and photons increases ion activity for Na^+/H^+ and Ca^{2+}/Na^+ antiporters as well as Ca^{2+} pumps and Na^+/K^+ ATPase. Furthermore, it regulates cAMP levels as well as Ca^{2+}, which controls various functions in the human body including signaling transfer, muscle contraction, blood coagulation, gene expression, etc. (Burnstock 2009; Gribble et al. 2000; Hamblin and Demidova 2006). Laser irradiation was found to trigger recurrent spikes in the intracellular calcium concentration (Alexandratou et al. 2002). The authors investigated the human fibroblast alterations induced by PBM at the single cell level using confocal microscopy. Their results showed that PBM causes temporary global Ca^{2+} release from intracellular stores as well as increase MMP, delta psi m, $\Delta\Psi m$. It is suggested that the increase in elevation of the intracellular Ca^{2+} concentration could be as result of influx of extracellular Ca^{2+} or from sequestered Ca^{2+} produced from the intracellular stores in the ER and mitochondria (Alexandratou et al. 2002).

4.2. Regulation of Transcription Factors

4.2.1. Nuclear Factor Kappa B (NF-κB)

NF-κB is a highly pleiotropic, redox sensitive nuclear transcription factor (Li and KARIN 1999; Tergaonkar 2006) that initiates many gene products known to be beneficial with PBM including the differentiation of myofibroblasts (Campbell and Perkins 2006). It is well known for its pro-inflammatory effect even though some studies have found it to be anti-inflammatory (De Lima et al. 2009). PBM is known for its inflammatory effect *in vivo*, however, studies have shown that PBM activates NF-κB. PBM activates NF-κB in mouse embryonic fibroblasts through ROS generation as well as through the phosphorylation of its inhibitor, IκB kinase (IKK) α/β which lead to its ubiquitination and degradation, thus allowing phosphorylation of the p65 subunit to prompt specific gene expression in the nucleus (Chen et al. 2009; Chen et al. 2011). Protein kinase is known to activate NF-kB in embryonic fibroblasts (Storz and Toker 2003). Lim et al.

(2013) studied the modulation of lipopolysaccharide-induced NF-κB signaling pathway with 635 nm via Heat Shock Protein 27 (HSP27) using human gingival fibroblast cells and found that PBM suppressed the release of prostaglandin E2 (PGE2), possibly through a mechanism related to the inhibition of NF-κB pathway. They suggested that the effect of PBM on NF-kB through HSP27 is necessary for down regulation of pro-inflammatory genes (Lim et al. 2013).

4.2.2. AKT/mTOR/Cyclin D1 Pathway

The protein kinase B/mammalian target of rapamycin (AKT/mTOR) signaling pathway are necessary for various biological functions including translation, control of cell cycle/growth, transcription, and cell proliferation (Hay and Sonenberg 2004; Ma and Blenis 2009). mTOR becomes activated through the phosphorylation of 4E binding protein 1 (4EBP1) and p70S6 kinase (p70S6k), and then modulates protein synthesis necessary for growth and metabolism (Parrales et al. 2013; Parrales et al. 2011). The dysregulation of mTOR is common in various human pathologies including cancers, neurological and cardiac disease and diabetes. Excessive stimulation of this pathway may enhance protein synthesis and cell growth. Phosphoinositide 3-kinases, P13K/AKT signaling pathway regulates mTOR upstream and plays a vital role in skin homeostasis as well as wound healing (Lima et al. 2012; Liu et al. 2019). However, the underlying mechanism is not clear especially with PBM. Xing et al. (2015) found that Acemannan promoted cyclin D1 expression in cultured fibroblast through the AKT/mTOR signaling pathway, which led to increase fibroblast proliferation and wound healing by regulation of cyclin D1 mRNA. PBM is known to modify the expression of proteins related to progression and invasion of cancer. This could aggravate the behavior of certain cancers increasing the expression of pAKT, pS6 and Cyclin D1 proteins (Sperandio et al. 2013).

4.2.3. Extracellular Regulated Protein Kinases/ Forkhead Box Transcription Factor (ERK/FOXM1)

Forkhead box M1 (FOXM1) and its isoforms are transcription factors or proto-oncogene involved in controlling cell cycle progression from G1 to S phase in various cell types including epithelial cells, dendritic cells, T and B cells or human cancers (Bella et al. 2014; Koo et al. 2012; Xue et al. 2010; Zhang et al. 2006). They control transcription of various genes such as cyclin B1 (*CCNB1*) and Serine/Threonine-Protein Kinase (*PLK1*) necessary for movement in G2/M, cyclin D1 (CCND1) for cell progression in G1/S as well as self-transcription. Also for the upregulation of Baculoviral IAP Repeat Containing 5 (BIRC5, Survivin) important in maintaining the integrity at the mitotic spindle checkpoint as well as anti-apoptotic activities (Barrett et al. 2011; Halasi and Gartel 2009; Wang et al. 2005). For the first time, Penke et al. (2018) identified FOXM1 as a driver of lung fibroblast proliferation. The authors found an upregulation of the expression of *FOXM1* and FOXM1 target genes in fibroblasts by overexpression of active AKT (Myr-AKT). With these, they concluded that PI3Kα signaling via PDK1/AKT mediates FGF2-induced *FOXM1* upregulation in lung fibroblasts (Penke et al. 2018). PBM activates fibroblast-signaling pathways such as the extracellular signal-regulated kinase (ERK)/FOXM1 pathway. Ling et al. (2014) used He-Ne laser (632.8 nm, 10 mW, 12.74 mW) to stimulate the extracellular regulated protein kinases (ERK) that was upstream of FOXM1 which led to FOXM1 phosphorylation and nuclear translocation. They also suggested that nuclear translocation improved FOXM1 transcriptional activity and its downstream target gene c-Mycexpression that could prevent p21 expression. PBM promoted FOXM1 nuclear translocation in an ERK-dependent manner, enhanced the transactivation of c-Myc, and ultimately resulted in inhibiting UVB-induced p21 expression. The results suggested that PBM can inhibit UVB-induced cell senescence through ERK/FOXM1 pathway. Furthermore, Shingyochi et al. (2017) showed that PBM (CO_2 laser, 10600 nm) activated AKT, ERK, and JNK to accelerate wound healing promoting migration and proliferation of fibroblasts which suggest the importance of ERK/FOXM1 in wound healing.

4.2.4. Peroxisome Proliferator-Activated Receptor Gamma (PPARγ)

Peroxisome proliferator-activated receptor (PPAR)-γ is a ligand-activated transcription factor belonging to the nuclear hormone receptor super family. They are expressed mainly in adipose tissue (Evans et al. 2004; Tyagi et al. 2011) but also in numerous immune cells including platelets, dendritic cells, monocytes/macrophages, and lymphocytes (Akbiyik et al. 2004; Asada et al. 2004; Padilla et al. 2002). It is made up of three domains including the DNA binding domain, agonist dependent activation domain (AF-1) and agonist-dependent activation domain (AF-2). PPAR-γ heterodimerizes with retinoid X receptor-α when bound to an agonist and initiates transcription response to target genes with the help of the PPAR response element (PPRE) (Auboeuf et al., 1997). It functions regulating metabolic activities of fatty acid production and glucose, generation of HSP-70, preventing inflammatory signal through NF-κB and promoting formation of adipose tissues. de Lima et al. (2013) found that PBM promotes homeostasis. They used PBM (660 nm, 5.4 J/cm^2) to irradiate the skin of rats over the bronchus. The authors noticed marked increase in HSP-70 and PPAR-y and concluded that the homeostatic function increases the expression of transcription factors initiating HSP70 production as well as other anti-inflammatory proteins. Investigations on the effects of PPAR in obesity, metabolism, and insulin sensitivity has made PPARγ agonists very prominent in treating type 2 diabetes however, it remains a therapeutic target for many inflammatory diseases (Croasdell et al. 2015).

4.2.5. *Runt-related Transcription Factor 2 (RUNX2)*

RUNX2, a member of the Runt family of transcription factors, and plays important roles in embryonic development to promote osteoblast differentiation and angiogenesis. Numerous pathways including PI3K, MAPK and STAT controls its activities (Fujita et al. 2004; Lee et al. 2002; Xiao et al. 2003). Furthermore, they also control several downstream target genes, such as growth factors and receptors extracellular matrix proteins, mitochondrial proteins, and transcription factors (Gaikwad, Cavender, and D'Souza 2001). PBM stimulates *in vitro* mineralization through increased insulin-like growth factor-I (IGF-I) and bone morphogenetic proteins (BMP) production, through Runx2 expression and ERK phosphorylation in osteoblasts (Kiyosaki et al. 2010). It increases RUNX2 in diabetic animals promoting a more appropriate tissue organization. Patrocínio-Silva et al.

(2014) observed increased RUNX2 expression using PBM at 830 nm to irradiate Wistar rats, which resulted to increase bone formation. Fibroblast growth factor (FGF)/FGF receptor (FGFR) signaling induces the expression of RUNX2, stimulating the DNA-binding and transcriptional activities of RUNX2 as well as its expression, and these are largely regulated by the protein kinase C (PKC) pathway (Kim et al. 2003)

4.3. Effector Molecules

4.3.1. Transforming Growth Factors (TGF-β)

Transforming growth factor-β1 (TGF-β1) is an important pro-inflammatory cytokine that is secreted during wound healing. It consist of various functions including controlling growth, proliferation, differentiation and apoptosis. TGF-β1 promotes collagen synthesis and decreased degradation during fibrogenesis in dermal fibroblast through the smad pathway (Ghahary et al. 1995; Hinz 2016; Hinz et al. 2007; Quan et al. 2010; Stahnke et al. 2017; Tredget et al. 2000). TGF-β1 is the main promoter of phenotypic shift between myofibroblasts and fibroblasts accompanied by the expression and incorporation of a-smooth muscle actin (a-SMA) (Hinz 2016; Taflinski et al. 2014). TGF-β1 is very vital in the wound healing process (Grande 1997; Schachtrup et al. 2010; Tandon et al. 2010). It promotes fibroblasts proliferation through mediated chemotaxis *in vivo*. It also recruits cells to the wound bed, synthesize ECM at the same time inhibiting proteases as well as stimulates growth and collagen secretion hence regarded as an integral factor to keloid formation. TGF-β1 also favors myofibroblasts trans-differentiation *in vitro* while α-SMA are been produced (Krummel et al. 1988; Vyas et al. 2010). To initiate smad2 and smad3 as well as other downstream signaling, several processes occur; the activation of its kinase, trans-phosphorylation of type I receptor (*TGFBRI*) and binding of the extracellular domain of its type II receptor (*TGFBRII*) (Attisano et al. 1994; Wrana et al. 1994).

Using molecular and pharmacologic approaches to inhibit FOXM1, differentiation of TGF-β–induced myofibroblast is also inhibited (Penke et al. 2018). PBM activates TGF-β in various fibroblast cells. Tang et al. (2017) found that PBM (810 nm) at low doses activated TGF in normal human oral fibroblasts. The authors also found that two arms of TGF-β1 signaling, Smad and non-Smad were involved in laser-mediated human β defensin-2 (HBD-2 or DEFB4A) expression. Hawkins and Abrahamse (2007) investigating the time-dependent responses of wounded human skin fibroblasts following phototherapy used two lasers namely Nd-YAG laser (1064 nm, 16 J/cm^2) and He-Ne (632.8 nm, 5/cm^2) and found that Nd-YAG laser reduced TGF-β secretion meanwhile He-Ne laser increased bFGF secretion. Furthermore, non-toxic doses of blue light significantly reduced the proliferation of TGF-β1 in HDF. After repeated irradiation, non-toxic doses significantly inhibits TGF-β1-induced differentiation of HDF into myofibroblasts (Taflinski et al. 2014). Nowak et al. (2000) demonstrated that using a super-pulsed CO$_2$ laser decreases TGF-β1 secretion and increases bFGF secretion in both normal and keloid dermal fibroblasts *in vitro*, promoting cell replication. This may assist in balancing the organization of collagen against fibrosis.

4.3.2. Oxidative Stress

Fibroblast cells may be subject to oxidative stress due to unfavorable cellular microenvironment characterized by an increased ROS production (Shatrova et al. 2016). Weng et al. (2011) investigated the irradiation of primary cultured skin fibroblasts by different non-ablative lasers to determine its effect on collagen synthesis and the antioxidant status, as well as identify a possible mechanism for laser photorejuvenation. Three different lasers were used namely: 532 nm potassium–titanyl phosphate (KTP), 1064 nm Q-switched neodymium:yttrium–aluminium–garnet (Nd:Yag) and 1064 nm long-pulse Nd:YAG. The study demonstrated an increase in collagen synthesis in all three lasers as well as two major antioxidant enzymes; superoxide dismutase (SOD) and glutathione peroxidase (GSH-Px). They concluded that improved antioxidant capacity reduces oxidative stress thereby increasing the synthesis of new collagen. Taflinski et al. (2014) demonstrated that higher doses of blue light (420 nm) induced intracellular oxidative stress in human dermal fibroblasts leading to toxicity. It was also observed that sub-toxic levels of oxidative stress was also observed when lower doses were utilized. The authors concluded that non-toxic irradiations using blue light might induce the energy consuming cellular responses against oxidative stress, which may interfere with myofibroblasts differentiation and increase cell metabolism.

4.3.3. Vascular Endothelial Growth Factor (VEGF)

Vascular endothelial growth factor (VEGF) functions in vascular remodeling as the main intermediary between fibroblasts and other cells. It promotes fibroblasts activation and specific ECM synthesis. Synthesis of VEGF is enhanced in all cell types by TGF-β1 (Larsson-Callerfelt et al. 2017). According to (Gkogkos et al. 2015) PBM (Nd:YAG 1064 nm, 2.6 to 15.8 J/cm^2) may induce gingival fibroblasts' proliferation, upregulate the secretion of EGF and promote increase in VEGF secretion. The authors suggested that repeated irradiations could possibly induce growth factor secretions and cell proliferation every 48 h. Another experimental study showed significant increase of VEGF secretion in myocardium fibroblasts when PBM (He-Ne 632 nm, 2.1 J/cm^2) was used (Kipshidze et al. 2001). Also Hakki and Bozkurt (2012) studied the effects of different setting of diode laser (940 nm) on the mRNA expression of growth factors and type I collagen of human gingival fibroblasts laser and found an increase in VEGF and TGF- β m RNA production.

4.3.4. Fibroblasts Growth Factor (bFGF)

Basic fibroblasts growth factor (bFGF) is potent mitogen that inhibits collagen production, stimulates proliferation, migration of cells including fibroblast and stabilizes cellular phenotype (Bennett et al. 2003). Shi et al. (2015) suggested that bFGF stimulates movement of diabetic human dermal fibroblasts *in vitro* through increased ROS production via the PI3K/AKT-Rac1-JNK pathways. PBM stimulates the expression of bFGF. After irradiation with diode laser (685 nm, 2 J/cm^2) there was a significant increase in the expression of bFGF (Saygun et al. 2008). In addition, Damante et al. (2009) found similar results using diode laser (660 nm, 3 J/cm^2 and 5 J/cm^2) in an *in vitro* study (Damante et al. 2009). bFGF expression was also increased in human skin fibroblasts using KTP laser (532 nm, 0.8 J/cm^2)

(Poon, Huang, and Burd 2005). In an experimental study with PBM, there was activation of interleukin-6 (IL-6)/bFGF mediated pathway Esmaeelinejad and Bayat (2013). The effect of PBM could also depend on the type of laser used. In studies with bFGF KTP and He-Ne has proven to be more effective. On the other hand, Usumez et al. (2014) in an animal study found that Nd:YAG (1064 nm) and diode laser therapy (980 nm, 8 J/cm^2) accelerated the wound healing process by changing the expression of PDGF and bFGF.

Conclusion

Even though, PBM is not widely accepted, its effect on various cells including fibroblasts is evidence of its therapeutic importance in reducing pain and inflammation. PBM modulates the mitochondrial respiratory chain generating small amounts of ROS which increases the MMP leading to increase ATP production. Ion channels are then activated allowing calcium to enter the cells and stimulate various signaling pathways leading to the activation of transcription factors which affect the protein expression, anti-inflammatory signaling, redox reactions, cell death mechanism, as well as stimulate processes to increase cell proliferation of various cell types, macrophage phagocytic activity, locomotion and maturation of fibroblasts. This decreases the inflammatory infiltrate, stimulating fibroblast proliferation and angiogenesis, and therefore recommended for wound healing processes.

Acknowledgment

This research was funded by South African Research Chairs Initiative of the Department of Science and Technology and National Research Foundation of South Africa (Grant no 98337).

REFERENCES

Agren, M. S. and Werthen, M. (2007). The extracellular matrix in wound healing: a closer look at therapeutics for chronic wounds. *The international journal of lower extremity wounds* 6 (2): 82-97.

Akbiyik, F., Ray, D. M., Gettings, K. F., Blumberg, N., Francis, C. W. and Phipps, R. P. (2004). Human bone marrow megakaryocytes and platelets express PPARγ, and PPARγ agonists blunt platelet release of CD40 ligand and thromboxanes. *Blood* 104 (5): 1361-1368.

Alberts, B., Johnson, A., Lewis, J., Raff, M., Roberts, K. and Walter, P. (2002). The extracellular matrix of animals. In *Molecular Biology of the Cell. 4th edition*. Garland Science.

Alexandratou, E., Yova, D., Handris, P., Kletsas, D. and Loukas, S. (2002). Human fibroblast alterations induced by low power laser irradiation at the single cell level using confocal microscopy. *Photochemical & Photobiological Sciences* 1 (8): 547-552. doi: 10.1039/B110213N.

Arany, P. R. 2012. Photobiomodulation: poised from the fringes. Mary Ann Liebert, Inc., New Rochelle, NY 10801 USA.

Asada, K., Sasaki, S., Suda, T., Chida, K. and Nakamura, H. (2004). Antiinflammatory roles of peroxisome proliferator–activated receptor γ in human alveolar macrophages. *American Journal of respiratory and critical care medicine* 169 (2): 195-200.

Ashcroft, K. J., Syed, F. and Bayat, A. (2013). Site-specific keloid fibroblasts alter the behaviour of normal skin and normal scar fibroblasts through paracrine signalling. *PloS one* 8 (12): e75600.

Assis, L., Moretti, A. I., Abrahão, T. B., Cury, V., Souza, H. P., Hamblin, M. R. and Parizotto, N. A. (2012). Low-level laser therapy (808 nm) reduces inflammatory response and oxidative stress in rat tibialis anterior muscle after cryolesion. *Lasers in surgery and medicine* 44 (9): 726-735.

Attisano, L., Wrana, J. L., López-Casillas, F. and Massagué, J. (1994). TGF-β receptors and actions. *Biochimica et Biophysica Acta (BBA)-Molecular Cell Research* 1222 (1): 71-80.

Badylak, S. F. (2002). "The extracellular matrix as a scaffold for tissue reconstruction." *Seminars in cell & developmental biology.* 13 (5): 377-383.

Bainbridge, P. (2013). Wound healing and the role of fibroblasts. *Journal of Wound Care* 22 (8): 407-408, 410-412. doi: 10.12968/jowc.2013.22.8.407.

Barrett, R. M., Colnaghi, R. and Wheatley, S. P. (2011). Threonine 48 in the BIR domain of survivin is critical to its mitotic and anti-apoptotic activities and can be phosphorylated by CK2 in vitro. *Cell Cycle* 10 (3): 538-548.

Barrientos, S., Stojadinovic, O., Golinko, M. S., Brem, H. and Tomic-Canic, M. (2008). Growth factors and cytokines in wound healing. *Wound repair and regeneration* 16 (5): 585-601.

Beldon, P. (2010). Basic science of wound healing. *Surgery (Oxford)* 28 (9): 409-412.

Bella, L., Zona, S., de Moraes, G. N. and Lam, E. W.-F. (2014). "FOXM1: a key oncofoetal transcription factor in health and disease." *Seminars in cancer biology.* 29: 32-39. https://doi.org/10.1016/j.semcancer.2014.07.008

Bennett, S. P., Griffiths, G. D., Schor, A. M., Leese, G. P. and Schor, S. L. (2003). Growth factors in the treatment of diabetic foot ulcers. *BJS* 90 (2): 133-146. doi: 10.1002/bjs.4019.

Bogaczewicz, J., Dudek, W., Wroński, J., Chodorowska, G., Przywara, S., Krasowska, D. and Zubilewicz, T. (2005). The role of matrix metalloproteinases in venous leg ulcers development. *Polski merkuriusz lekarski: organ Polskiego Towarzystwa Lekarskiego* 19 (113): 686-692.

Bohr, N. (1921). Atomic structure. *Nature* 107 (2682): 104.

Bos, J. L. (2003). Epac: a new cAMP target and new avenues in cAMP research. *Nature reviews Molecular cell biology* 4 (9): 733.

Burnstock, G. (2009). Purines and sensory nerves. In *Sensory Nerves*, 333-392. Springer.

Calabrese, E. J. and Mattson, M. P. (2017). How does hormesis impact biology, toxicology, and medicine? *NPJ aging and mechanisms of disease* 3: 13-13. doi: 10.1038/s41514-017-0013-z.

Campbell, K. J. and Perkins, N. D. 2006. "Regulation of NF-κB function." *Biochemical Society Symposia*.

Campos, A. C., Groth, A. K. and Branco, A. B. (2008). Assessment and nutritional aspects of wound healing. *Current Opinion in Clinical Nutrition & Metabolic Care* 11 (3): 281-288.

Chance, B., Cooper, C. E., Delpy, D. T., Reynolds, E. O. R., Delpy, D. T. and Cope, M. (1997). Quantification in tissue near–infrared spectroscopy. *Philosophical Transactions of the Royal Society of London. Series B: Biological Sciences* 352 (1354): 649-659. doi: 10.1098/rstb.1997.0046.

Chaves, M. E. d. A., Araújo, A. R. d., Piancastelli, A. C. C. and Pinotti, M. (2014). Effects of low-power light therapy on wound healing: LASER x LED. *Anais brasileiros de dermatologia* 89 (4): 616-623.

Chen, A. C.-H., Huang, Y.-Y., Arany, P. R. and Hamblin, M. R. 2009. *Role of reactive oxygen species in low level light therapy.* Vol. 7165, *SPIE BiOS*: SPIE.

Chen, A. C. H., Arany, P. R., Huang, Y.-Y., Tomkinson, E. M., Sharma, S. K., Kharkwal, G. B., Saleem, T., Mooney, D., Yull, F. E., Blackwell, T. S. and Hamblin, M. R. (2011). Low-level laser therapy activates NF-kB via generation of reactive oxygen species in mouse embryonic fibroblasts. *PloS one* 6 (7): e22453-e22453. doi: 10.1371/journal.pone.0022453.

Chen, Y., Corriden, R., Inoue, Y., Yip, L., Hashiguchi, N., Zinkernagel, A., Nizet, V., Insel, P. A. and Junger, W. G. (2006). ATP release guides neutrophil chemotaxis via P2Y2 and A3 receptors. *Science* 314 (5806): 1792-1795.

Chen, Y. J., Jeng, J. H., Lee, B. S., Chang, H. F., Chen, K. C. and Lan, W. H. (2000). Effects of Nd:YAG laser irradiation on cultured human gingival fibroblasts. *Lasers in Surgery and Medicine* 27 (5): 471-478. doi: 10.1002/1096-9101(2000)27:5<471::aid-lsm1008>3.0.co;2-q.

Chrysanthopoulou, A., Mitroulis, I., Apostolidou, E., Arelaki, S., Mikroulis, D., Konstantinidis, T., Sivridis, E., Koffa, M., Giatromanolaki, A., Boumpas, D. T., Ritis, K. and Kambas, K. (2014). Neutrophil extracellular traps promote differentiation and function of fibroblasts. *The Journal of Pathology* 233 (3): 294-307. doi: 10.1002/path.4359.

Chung, H., Dai, T., Sharma, S. K., Huang, Y.-Y., Carroll, J. D. and Hamblin, M. R. (2012). The nuts and bolts of low-level laser (light) therapy. *Annals of biomedical engineering* 40 (2): 516-533. doi: doi: 10.1007/s10439-011-0454-7.

Cook, H., Stephens, P., Davies, K. J., Thomas, D. W. and Harding, K. G. (2000). Defective extracellular matrix reorganization by chronic wound fibroblasts is associated with alterations in TIMP-1, TIMP-2, and MMP-2 activity. *Journal of Investigative Dermatology* 115 (2): 225-233.

Croasdell, A., Duffney, P. F., Kim, N., Lacy, S. H., Sime, P. J. and Phipps, R. P. (2015). PPARγ and the Innate Immune System Mediate the Resolution of Inflammation. *PPAR Research* 2015: 20. doi: 10.1155/2015/549691.

Damante, C. A., De Micheli, G., Miyagi, S. P. H., Feist, I. S. and Marques, M. M. (2009). Effect of laser phototherapy on the release of fibroblast growth factors by human gingival fibroblasts. *Lasers in Medical Science* 24 (6): 885.

Dawood, M. S. and Salman, S. D. (2013). Low level diode laser accelerates wound healing. *Lasers in medical science* 28 (3): 941-945. doi: https://doi.org/10.1007/s10103-012-1182-4.

de Lima, F. M., Albertini, R., Dantas, Y., Maia-Filho, A. L., de Loura Santana, C., Castro-Faria-Neto, H. C., França, C., Villaverde, A. B. and Aimbire, F. (2013). Low-level laser therapy restores the oxidative stress balance in acute lung injury induced by gut ischemia and reperfusion. *Photochemistry and photobiology* 89 (1): 179-188.

De Lima, F. M., Naves, K., Machado, A., Albertini, R., Villaverde, A. and Aimbire, F. (2009). Lung inflammation and endothelial cell damage are decreased after treatment with phototherapy (PhT) in a model of acute lung injury induced by Escherichia coli lipopolysaccharide in the rat. *Cell biology international* 33 (12): 1212-1221.

Desmoulière, A., Chaponnier, C. and Gabbiani, G. (2005). Tissue repair, contraction, and the myofibroblast. *Wound repair and regeneration* 13 (1): 7-12.

Driskell, R. R., Lichtenberger, B. M., Hoste, E., Kretzschmar, K., Simons, B. D., Charalambous, M., Ferron, S. R., Herault, Y., Pavlovic, G. and Ferguson-Smith, A. C. (2013). Distinct fibroblast lineages determine dermal architecture in skin development and repair. *Nature* 504 (7479): 277.

Eells, J. T., Henry, M., Summerfelt, P., Wong-Riley, M., Buchmann, E., Kane, M., Whelan, N. and Whelan, H. (2003). Therapeutic photobiomodulation for methanol-induced retinal toxicity. *Proceedings of the National Academy of Sciences* 100 (6): 3439-3444.

Einstein, A. (1917). Zur quantentheorie der strahlung. *Phys. Z.* 18: 121-128.

El Ghalbzouri, A., Hensbergen, P., Gibbs, S., Kempenaar, J., van der Schors, R. and Ponec, M. (2004). Fibroblasts facilitate re-epithelialization in wounded human skin equivalents. *Laboratory investigation* 84 (1): 102.

Elliott, M. R., Chekeni, F. B., Trampont, P. C., Lazarowski, E. R., Kadl, A., Walk, S. F., Park, D., Woodson, R. I., Ostankovich, M. and Sharma, P. (2009). Nucleotides released by apoptotic cells act as a find-me signal to promote phagocytic clearance. *Nature* 461 (7261): 282.

Enoch, S. and Leaper, D. J. (2008). Basic science of wound healing. *Surgery (Oxford)* 26 (2): 31-37.

Esmaeelinejad, M. and Bayat, M. (2013). Effect of low-level laser therapy on the release of interleukin-6 and basic fibroblast growth factor from cultured human skin fibroblasts in normal and high glucose mediums. *Journal of Cosmetic and Laser Therapy* 15 (6): 310-317.

Evans, D. H. and Abrahamse, H. (2009). A review of laboratory-based methods to investigate second messengers in low-level laser therapy (LLLT). *Medical Laser Application* 24 (3): 201-215.

Evans, R. M., Barish, G. D. and Wang, Y.-X. (2004). PPARs and the complex journey to obesity. *Nature Medicine* 10 (4): 355-361. doi: 10.1038/nm1025.

Falanga, V. (2005). Wound healing and its impairment in the diabetic foot. *The Lancet* 366 (9498): 1736-1743.

Fernandes, A., Neto, N. L., Marques, N. C. T., Vitor, L. L. R., Prado, M. T. O., Oliveira, R. C., Machado, M. A. A. M. and Oliveira, T. M. (2018). Cellular response of pulp fibroblast to single or multiple photobiomodulation applications. *Laser Physics* 28 (6): 065604. doi: 10.1088/1555-6611/aab952.

Freitas, L. F. d. and Hamblin, M. R. (2016). Proposed Mechanisms of Photobiomodulation or Low-Level Light Therapy. *IEEE Journal of Selected Topics in Quantum Electronics* 22 (3): 348-364. doi: 10.1109/JSTQE.2016.2561201.

Fujita, T., Azuma, Y., Fukuyama, R., Hattori, Y., Yoshida, C., Koida, M., Ogita, K. and Komori, T. (2004). Runx2 induces osteoblast and chondrocyte differentiation and enhances their migration by coupling with PI3K-Akt signaling. *Journal of Cell Biology* 166 (1): 85-95.

Gaikwad, J. S., Cavender, A. and D'Souza, R. N. (2001). Identification of tooth-specific downstream targets of Runx2. *Gene* 279 (1): 91-97.

George, B., II, Janis, J. E. and Attinger, C. E. (2006). The basic science of wound healing. *Plastic and reconstructive surgery* 117 (7S): 12S-34S.

George, S., Hamblin, M. R. and Abrahamse, H. (2018). Effect of red light and near infrared laser on the generation of reactive oxygen species in primary dermal fibroblasts. *Journal of Photochemistry and Photobiology B: Biology* 188: 60-68. doi: https://doi.org/10.1016/j.jphotobiol.2018.09.004.

Gerasimovskaya, E. V., Ahmad, S., White, C. W., Jones, P. L., Carpenter, T. C. and Stenmark, K. R. (2002). Extracellular ATP Is an Autocrine/Paracrine Regulator of Hypoxia-induced Adventitial Fibroblast Growth: Signaling through Extracellular Signal-Regulated Kinase-1/2 and the Egr-1 Transcription Factor. *Journal of Biological Chemistry* 277 (47): 44638-44650. doi: 10.1074/jbc.M203012200.

Ghahary, A., Shen, Y. S., Scott, P. G. and Tredget, E. E. (1995). Immunolocalization of TGF-β1 in human hypertrophic scar and normal dermal tissues. *Cytokine* 7 (2): 184-190.

Gkogkos, A. S., Karoussis, I. K., Prevezanos, I. D., Marcopoulou, K. E., Kyriakidou, K. and Vrotsos, I. A. (2015). Effect of Nd:YAG Low Level Laser Therapy on Human Gingival Fibroblasts. *International Journal of Dentistry* 2015: 7. doi: 10.1155/2015/258941.

Gordon, J. P., Zeiger, H. J. and Townes, C. H. (1955). The maser—new type of microwave amplifier, frequency standard, and spectrometer. *Physical Review* 99 (4): 1264.

Gosain, A. and DiPietro, L. A. (2004). Aging and wound healing. *World journal of surgery* 28 (3): 321-326.

Grande, J. P. (1997). Role of transforming growth factor-β in tissue injury and repair. *Proceedings of the Society for Experimental Biology and Medicine* 214 (1): 27-40.

Gribble, F. M., Loussouarn, G., Tucker, S. J., Zhao, C., Nichols, C. G. and Ashcroft, F. M. (2000). A novel method for measurement of submembrane ATP concentration. *Journal of Biological Chemistry* 275 (39): 30046-30049.

Guo, S. a. and DiPietro, L. A. (2010). Factors affecting wound healing. *Journal of dental research* 89 (3): 219-229.

Gurtner, G. C., Werner, S., Barrandon, Y. and Longaker, M. T. (2008). Wound repair and regeneration. *Nature* 453 (7193): 314.

Hakki, S. S. and Bozkurt, S. B. (2012). Effects of different setting of diode laser on the mRNA expression of growth factors and type I collagen of human gingival fibroblasts. *Lasers in medical science* 27 (2): 325-331.

Halasi, M. and Gartel, A. L. (2009). A novel mode of FoxM1 regulation: positive auto-regulatory loop. *Cell Cycle* 8 (12): 1966-1967.

Halayko, A. J., Tran, T. and Gosens, R. (2008). Phenotype and functional plasticity of airway smooth muscle: role of caveolae and caveolins. *Proceedings of the American Thoracic Society* 5 (1): 80-88.

Hamblin, M. R. (2018). Photobiomodulation for traumatic brain injury and stroke. *Journal of neuroscience research* 96 (4): 731-743.

Hamblin, M. R. and Demidova, T. N. 2006. *Mechanisms of low level light therapy*. Vol. 6140, *SPIE BiOS*: SPIE.

Hawkins, D. H. and Abrahamse, H. (2006). The role of laser fluence in cell viability, proliferation, and membrane integrity of wounded human skin fibroblasts following helium-neon laser irradiation. *Lasers in Surgery and Medicine* 38 (1): 74-83. doi: 10.1002/lsm.20271.

Hawkins, D. H. and Abrahamse, H. (2007). Time-dependent responses of wounded human skin fibroblasts following phototherapy. *Journal of Photochemistry and Photobiology B: Biology* 88 (2-3): 147-155.

Hay, N. and Sonenberg, N. (2004). Upstream and downstream of mTOR. *Genes & development* 18 (16): 1926-1945.

Hayashi, T., Lin, I.-J., Chen, Y., Fee, J. A. and Moënne-Loccoz, P. (2007). Fourier Transform Infrared Characterization of a CuB− Nitrosyl Complex in Cytochrome ba 3 from Thermus thermophilus: Relevance to NO Reductase Activity in Heme− Copper Terminal Oxidases. *Journal of the American Chemical Society* 129 (48): 14952-14958.

Herouy, Y., Trefzer, D., Zimpfer, U., Schöpf, E., Vanscheidt, W. and Norgauer, J. (2000). Matrix metalloproteinases and venous leg ulceration. *European Journal of Dermatology* 10 (3): 173-180.

Hinz, B. (2016). The role of myofibroblasts in wound healing. *Current Research in Translational Medicine* 64 (4): 171-177. doi: https://doi.org/10.1016/j.retram.2016.09.003.

Hinz, B., Phan, S. H., Thannickal, V. J., Galli, A., Bochaton-Piallat, M.-L. and Gabbiani, G. (2007). The myofibroblast: one function, multiple origins. *The American journal of pathology* 170 (6): 1807-1816.

Houreld, N. N., Masha, R. T. and Abrahamse, H. (2012). Low-intensity laser irradiation at 660 nm stimulates cytochrome c oxidase in stressed fibroblast cells. *Lasers in Surgery and Medicine* 44 (5): 429-434. doi: 10.1002/lsm.22027.

Huang, Y.-Y., Chen, A. C. H., Carroll, J. D. and Hamblin, M. R. (2009). Biphasic Dose Response in Low Level Light Therapy. *Dose-Response* 7 (4): dose-response.09-027.Hamblin. doi: 10.2203/dose-response.09-027.Hamblin.

Huang, Y.-Y., Sharma, S. K., Carroll, J. and Hamblin, M. R. (2011). Biphasic dose response in low level light therapy–an update. *Dose-Response* 9 (4): dose-response. 11-009. Hamblin.

Janssen, L. J., Farkas, L., Rahman, T. and Kolb, M. R. (2009). ATP stimulates Ca^{2+}-waves and gene expression in cultured human pulmonary fibroblasts. *The international journal of biochemistry & cell biology* 41 (12): 2477-2484.

Javan, A., Bennett Jr, W. R. and Herriott, D. R. (1961). Population inversion and continuous optical maser oscillation in a gas discharge containing a He-Ne mixture. *Physical Review Letters* 6 (3): 106.

Jenkins, P. A. and Carroll, J. D. (2011). How to report low-level laser therapy (LLLT)/photomedicine dose and beam parameters in clinical and laboratory studies. *Photomedicine and laser surgery* 29 (12): 785-787.

Johnstone, D. M., el Massri, N., Moro, C., Spana, S., Wang, X. S., Torres, N., Chabrol, C., De Jaeger, X., Reinhart, F., Purushothuman, S., Benabid, A. L., Stone, J. and Mitrofanis, J. (2014). Indirect application of near infrared light induces neuroprotection in a mouse model of parkinsonism – An abscopal neuroprotective effect. *Neuroscience* 274: 93-101. doi: //0-dx.doi.org.ujlink.uj.ac.za/10.1016/j.neuroscience.2014.05.023.

Karu, T. (1989). Laser biostimulation: a photobiological phenomenon. *Journal of Photochemistry and Photobiology B: Biology* 3 (4): 638.

Karu, T. 2010a. Mitochondrial mechanisms of photobiomodulation in context of new data about multiple roles of ATP. Mary Ann Liebert, Inc. 140 Huguenot Street, 3rd Floor New Rochelle, NY 10801 USA.

Karu, T. and Afanas'eva, N. (1995). "Cytochrome C oxidase as the primary photoacceptor upon laser exposure of cultured cells to visible and near IR-range light." Doklady Akademii Nauk.

Karu, T. I. (2010b). Multiple roles of cytochrome C oxidase in mammalian cells under action of red and IR-A radiation. *IUBMB Life* 62 (8): 607-610. doi: 10.1002/iub.359.

Karu, T. I. and Kolyakov, S. (2005). Exact action spectra for cellular responses relevant to phototherapy. *Photomedicine and Laser Therapy* 23 (4): 355-361.

Karu, T. I., Pyatibrat, L. V., Kolyakov, S. F. and Afanasyeva, N. I. (2008). Absorption measurements of cell monolayers relevant to mechanisms of laser phototherapy: reduction or oxidation of cytochrome c oxidase under laser radiation at 632.8 nm. *Photomedicine and laser surgery* 26 (6): 593-599.

Khan, I. and Arany, P. (2015). Biophysical approaches for oral wound healing: emphasis on photobiomodulation. *Advances in wound care* 4 (12): 724-737.

Kim, H.-J., Kim, J.-H., Bae, S.-C., Choi, J.-Y., Kim, H.-J. and Ryoo, H.-M. (2003). The Protein Kinase C Pathway Plays a Central Role in the Fibroblast Growth Factor-stimulated Expression and Transactivation Activity of Runx2. *Journal of Biological Chemistry* 278 (1): 319-326. doi: 10.1074/jbc.M203750200.

Kipshidze, N., Nikolaychik, V., Keelan, M. H., Shankar, L. R., Khanna, A., Kornowski, R., Leon, M. and Moses, J. (2001). Low-power helium: Neon laser irradiation enhances production of vascular endothelial growth factor and promotes growth of endothelial cells in vitro. *Lasers in Surgery and Medicine: The Official Journal of the American Society for Laser Medicine and Surgery* 28 (4): 355-364.

Kiyosaki, T., Mitsui, N., Suzuki, N. and Shimizu, N. (2010). Low-Level Laser Therapy Stimulates Mineralization Via Increased Runx2 Expression and ERK Phosphorylation in Osteoblasts. *Photomedicine and Laser Surgery* 28 (S1): S-167-S-172. doi: 10.1089/pho.2009.2693.

Koo, C.-Y., Muir, K. W. and Lam, E. W.-F. (2012). FOXM1: From cancer initiation to progression and treatment. *Biochimica et Biophysica Acta (BBA)-Gene Regulatory Mechanisms* 1819 (1): 28-37.

Kov, I. B., Mester, E. and Görög, P. (1974). Stimulation of wound healing with laser beam in the rat. *Experientia* 30 (11): 1275-1276.

Krassovka, J., Borgschulze, A., Sahlender, B., Lögters, T., Windolf, J. and Grotheer, V. (2019). Blue light irradiation and its beneficial effect on Dupuytren's fibroblasts. *PLOS ONE* 14 (1): e0209833. doi: 10.1371/journal.pone.0209833.

Krummel, T. M., Michna, B. A., Thomas, B. L., Sporn, M. B., Nelson, J. M., Salzberg, A. M., Cohen, I. K. and Diegelmann, R. F. (1988). Transforming growth factor beta (TGF-β) induces fibrosis in a fetal wound model. *Journal of pediatric surgery* 23 (7): 647-652.

Larsson-Callerfelt, A.-K., Andersson Sjöland, A., Hallgren, O., Bagher, M., Thiman, L., Löfdahl, C.-G., Bjermer, L. and Westergren-Thorsson, G. (2017). VEGF induces ECM synthesis and fibroblast activity in human lung fibroblasts. *European Respiratory Journal* 50 (suppl 61): PA1045. doi: 10.1183/1393003.congress-2017.PA1045.

Lee, K.-S., Hong, S.-H. and Bae, S.-C. (2002). Both the Smad and p38 MAPK pathways play a crucial role in Runx2 expression following induction by transforming growth factor-β and bone morphogenetic protein. *Oncogene* 21 (47): 7156.

Li, J., Chen, J. and Kirsner, R. (2007). Pathophysiology of acute wound healing. *Clinics in dermatology* 25 (1): 9-18.

Li, N. and KARIN, M. (1999). Is NF-κB the sensor of oxidative stress? *The FASEB journal* 13 (10): 1137-1143.

Liebert, A. D., Bicknell, B. T. and Adams, R. D. (2014). Protein conformational modulation by photons: A mechanism for laser treatment effects. *Medical hypotheses* 82 (3): 275-281.

Lim, W., Kim, J., Kim, S., Karna, S., Won, J., Jeon, S. M., Kim, S. Y., Choi, Y., Choi, H. and Kim, O. (2013). Modulation of Lipopolysaccharide-Induced NF-κB Signaling Pathway by 635 nm Irradiation via Heat Shock Protein 27 in Human Gingival Fibroblast Cells. *Photochemistry and Photobiology* 89 (1): 199-207. doi: 10.1111/j.1751-1097.2012.01225.x.

Lima, M. H., Caricilli, A. M., de Abreu, L. L., Araújo, E. P., Pelegrinelli, F. F., Thirone, A. C., Tsukumo, D. M., Pessoa, A. F. M., dos Santos, M. F. and de Moraes, M. A. (2012). Topical insulin accelerates wound healing in diabetes by enhancing the AKT and ERK pathways: a double-blind placebo-controlled clinical trial. *PloS one* 7 (5): e36974.

Lima, P. L. V., Pereira, C. V., Nissanka, N., Arguello, T., Gavini, G., Maranduba, C. M. d. C., Diaz, F. and Moraes, C. T. (2019). Photobiomodulation enhancement of cell proliferation at 660 nm does not require cytochrome c oxidase. *Journal of Photochemistry and Photobiology B: Biology* 194: 71-75. doi: https://doi.org/10.1016/j.jphotobiol.2019.03.015.

Ling, Q., Meng, C., Chen, Q. and Xing, D. (2014). Activated ERK/FOXM1 Pathway by Low-Power Laser Irradiation Inhibits UVB-Induced Senescence Through Down-Regulating p21 Expression. *Journal of Cellular Physiology* 229 (1): 108-116. doi: 10.1002/jcp.24425.

Liu, P., Choi, J.-W., Lee, M.-K., Choi, Y.-H. and Nam, T.-J. (2019). Wound Healing Potential of Spirulina Protein on CCD-986sk Cells. *Marine Drugs* 17 (2): 130.

Lohman, A. W., Billaud, M. and Isakson, B. E. (2012). Mechanisms of ATP release and signalling in the blood vessel wall. *Cardiovascular research* 95 (3): 269-280.

Lorenz, H. P. and Longaker, M. T. (2008). Wounds: biology, pathology, and management. In *Surgery*, 191-208. Springer.

Ma, X. M. and Blenis, J. (2009). Molecular mechanisms of mTOR-mediated translational control. *Nature Reviews Molecular Cell Biology* 10: 307. doi: 10.1038/nrm2672.

Maiman, T. H. (1960). Stimulated optical radiation in ruby.

Marques, M. M., Pereira, A. N., Fujihara, N. A., Nogueira, F. N. and Eduardo, C. P. (2004). Effect of low-power laser irradiation on protein synthesis and ultrastructure of human gingival fibroblasts. *Lasers in surgery and medicine* 34 (3): 260-265.

Martius, F. (1923). Das amdt-schulz grandgesetz. *Munch Med Wschr* 70: 1005-1006.

Mathieu, D., Linke, J.-C. and Wattel, F. (2006). Non-healing wounds. In *Handbook on hyperbaric medicine*, 401-428. Springer.

Mester, E., Mester, A. F. and Mester, A. (1985). The biomedical effects of laser application. *Lasers in Surgery and Medicine* 5 (1): 31-39. doi: 10.1002/lsm.1900050105.

Mester, E., Szende, B. and Gärtner, P. (1968). The effect of laser beams on the growth of hair in mice. *Radiobiologia, radiotherapia* 9 (5): 621-626.

Mester, E., Szende, B. and Tota, J. (1967). Effect of laser on hair growth of mice. *Kiserl Orvostud* 19: 628-631.

Mirastschijski, U., Haaksma, C. J., Tomasek, J. J. and Ågren, M. S. (2004). Matrix metalloproteinase inhibitor GM 6001 attenuates keratinocyte migration, contraction and myofibroblast formation in skin wounds. *Experimental cell research* 299 (2): 465-475.

Montminy, M. (1997). Transcriptional Regulation by Cyclic AMP. *Annual Review of Biochemistry* 66 (1): 807-822. doi: 10.1146/annurev.biochem.66.1.807.

Moore, P., Ridgway, T. D., Higbee, R. G., Howard, E. W. and Lucroy, M. D. (2005). Effect of wavelength on low-intensity laser irradiation-stimulated cell proliferation in vitro. *Lasers in Surgery and Medicine* 36 (1): 8-12. doi: 10.1002/lsm.20117.

Navarro-Requena, C., Pérez-Amodio, S., Castaño, O. and Engel, E. (2018). Wound healing-promoting effects stimulated by extracellular calcium and calcium-releasing nanoparticles on dermal fibroblasts. *Nanotechnology* 29 (39): 395102. doi: 10.1088/1361-6528/aad01f.

Newton, P., Watson, J., Wolowacz, R. and Wood, E. (2004). Macrophages restrain contraction of an in vitro wound healing model. *Inflammation* 28 (4): 207-214.

Nowak, K. C., McCormack, M. and Koch, R. J. (2000). The Effect of Superpulsed Carbon Dioxide Laser Energy on Keloid and Normal Dermal Fibroblast Secretion of Growth Factors: A Serum-Free Study. *Plastic and Reconstructive Surgery* 105 (6): 2039-2048.

Padilla, J., Leung, E. and Phipps, R. P. (2002). Human B lymphocytes and B lymphomas express PPAR-γ and are killed by PPAR-γ agonists. *Clinical Immunology* 103 (1): 22-33.

Parrales, A., López, E., Lee-Rivera, I. and López-Colomé, A. M. (2013). ERK1/2-dependent activation of mTOR/mTORC1/p70S6K regulates thrombin-induced RPE cell proliferation. *Cellular signalling* 25 (4): 829-838.

Parrales, A., Lopez, E. and Lopez-Colome, A. (2011). Thrombin activation of PI3K/PDK1/Akt signaling promotes cyclin D1 upregulation and RPE cell proliferation. *Biochimica et Biophysica Acta (BBA)-Molecular Cell Research* 1813 (10): 1758-1766.

Patrocínio-Silva, T. L., de Souza, A. M. F., Goulart, R. L., Pegorari, C. F., Oliveira, J. R., Fernandes, K., Magri, A., Pereira, R. M. R., Araki, D. R. and Nagaoka, M. R. (2014). The effects of low-level laser irradiation on bone tissue in diabetic rats. *Lasers in medical science* 29 (4): 1357-1364.

Penke, L. R., Speth, J. M., Dommeti, V. L., White, E. S., Bergin, I. L. and Peters-Golden, M. (2018). FOXM1 is a critical driver of lung fibroblast activation and fibrogenesis. *The Journal of clinical investigation* 128 (6): 2389-2405. doi: 10.1172/JCI87631.

Penn, J. W., Grobbelaar, A. O. and Rolfe, K. J. (2012). The role of the TGF-β family in wound healing, burns and scarring: a review. *International journal of burns and trauma* 2 (1): 18.

Phan, S. H. (2008). Biology of fibroblasts and myofibroblasts. *Proceedings of the American Thoracic Society* 5 (3): 334-337.

Pinheiro, A. R., Paramos-de-Carvalho, D., Certal, M., Costa, M. A., Costa, C., Magalhães-Cardoso, M. T., Ferreirinha, F., Sévigny, J. and Correia-de-Sá, P. (2013). Histamine induces ATP release from human subcutaneous fibroblasts, via pannexin-1 hemichannels, leading to Ca2+ mobilization and cell proliferation. *Journal of Biological Chemistry* 288 (38): 27571-27583.

Plikus, M. V., Guerrero-Juarez, C. F., Ito, M., Li, Y. R., Dedhia, P. H., Zheng, Y., Shao, M., Gay, D. L., Ramos, R. and Hsi, T.-C. (2017). Regeneration of fat cells from myofibroblasts during wound healing. *Science* 355 (6326): 748-752.

Poon, V. K. M., Huang, L. and Burd, A. (2005). Biostimulation of dermal fibroblast by sublethal Q-switched Nd:YAG 532nm laser: Collagen remodeling and pigmentation. *Journal of Photochemistry and Photobiology B: Biology* 81 (1): 1-8. doi: https://doi.org/10.1016/j.jphotobiol.2005.05.006.

Pourzarandian, A., Watanabe, H., Ruwanpura, S. M. P. M., Aoki, A. and Ishikawa, I. (2005). Effect of Low-Level Er:YAG Laser Irradiation on Cultured Human Gingival Fibroblasts. *Journal of Periodontology* 76 (2): 187-193. doi: 10.1902/jop.2005.76.2.187.

Poyton, R. O. and Ball, K. A. (2011). Therapeutic photobiomodulation: nitric oxide and a novel function of mitochondrial cytochrome c oxidase. *Discovery medicine* 11 (57): 154-159.

Quan, T., Shao, Y., He, T., Voorhees, J. J. and Fisher, G. J. (2010). Reduced expression of connective tissue growth factor (CTGF/CCN2) mediates collagen loss in chronologically aged human skin. *The Journal of investigative dermatology* 130 (2): 415-424. doi: 10.1038/jid.2009.224.

Quan, T., Wang, F., Shao, Y., Rittié, L., Xia, W., Orringer, J. S., Voorhees, J. J. and Fisher, G. J. (2013). Enhancing Structural Support of the Dermal Microenvironment Activates Fibroblasts, Endothelial Cells, and

Keratinocytes in Aged Human Skin In Vivo. *Journal of Investigative Dermatology* 133 (3): 658-667. doi: 10.1038/jid.2012.364.

Rasik, A. M. and Shukla, A. (2000). Antioxidant status in delayed healing type of wounds. *International journal of experimental pathology* 81 (4): 257-263.

Rauch, U. (2004). Extracellular matrix components associated with remodeling processes in brain. *Cellular and Molecular Life Sciences CMLS* 61 (16): 2031-2045.

Rezende, S. B., Ribeiro, M. S., Nunez, S. C., Garcia, V. G. and Maldonado, E. P. (2007). Effects of a single near-infrared laser treatment on cutaneous wound healing: biometrical and histological study in rats. *Journal of Photochemistry and Photobiology B: Biology* 87 (3): 145-153.

Rhee, S. (2009). Fibroblasts in three dimensional matrices: cell migration and matrix remodeling. *Experimental & molecular medicine* 41 (12): 858.

Rognoni, E., Pisco, A. O., Hiratsuka, T., Sipilä, K. H., Belmonte, J. M., Mobasseri, S. A., Philippeos, C., Dilão, R. and Watt, F. M. (2018). Fibroblast state switching orchestrates dermal maturation and wound healing. *Molecular Systems Biology* 14 (8): e8174. doi: 10.15252/msb.20178174.

Rojas, J. C. and Gonzalez-Lima, F. (2013). Neurological and psychological applications of transcranial lasers and LEDs. *Biochemical pharmacology* 86 (4): 447-457. doi: //0-dx.doi.org.ujlink.uj.ac.za/10.1016/j.bcp.2013.06.012.

Rolfe, K. J. and Grobbelaar, A. O. (2010). The growth receptors and their role in wound healing. *Current opinion in investigational drugs (London, England: 2000)* 11 (11): 1221-1228.

Sáez, J. C., Berthoud, V. M., Branes, M. C., Martinez, A. D. and Beyer, E. C. (2003). Plasma membrane channels formed by connexins: their regulation and functions. *Physiological reviews* 83 (4): 1359-1400.

Santoro, M. M. and Gaudino, G. (2005). Cellular and molecular facets of keratinocyte reepithelization during wound healing. *Experimental cell research* 304 (1): 274-286.

Saygun, I., Karacay, S., Serdar, M., Ural, A. U., Sencimen, M. and Kurtis, B. (2008). Effects of laser irradiation on the release of basic fibroblast growth factor (bFGF), insulin like growth factor-1 (IGF-1), and receptor of IGF-1 (IGFBP3) from gingival fibroblasts. *Lasers in Medical Science* 23 (2): 211-215.

Schachtrup, C., Ryu, J. K., Helmrick, M. J., Vagena, E., Galanakis, D. K., Degen, J. L., Margolis, R. U. and Akassoglou, K. (2010). Fibrinogen triggers astrocyte scar formation by promoting the availability of active TGF-β after vascular damage. *Journal of Neuroscience* 30 (17): 5843-5854.

Schawlow, A. L. and Townes, C. H. (1958). Infrared and optical masers. *Physical Review* 112 (6): 1940.

Schiller, M., Dennler, S., Anderegg, U., Kokot, A., Simon, J. C., Luger, T. A., Mauviel, A. and Böhm, M. (2010). Increased cAMP levels modulate transforming growth factor-β/smad-induced expression of extracellular matrix components and other key fibroblast effector functions. *Journal of Biological Chemistry* 285 (1): 409-421.

Schwiebert, E. M. and Zsembery, A. (2003). Extracellular ATP as a signaling molecule for epithelial cells. *Biochimica et Biophysica Acta (BBA)-Biomembranes* 1615 (1-2): 7-32.

Shatrova, A. N., Lyublinskaya, O. G., Borodkina, A. V. and Burova, E. B. (2016). Oxidative stress response of human fibroblasts and endometrial mesenchymal stem cells. *Cell and Tissue Biology* 10 (1): 18-28. doi: 10.1134/s1990519x16010090.

Shi, H., Cheng, Y., Ye, J., Cai, P., Zhang, J., Li, R., Yang, Y., Wang, Z., Zhang, H., Lin, C., Lu, X., Jiang, L., Hu, A., Zhu, X., Zeng, Q., Fu, X., Li, X. and Xiao, J. (2015). bFGF Promotes the Migration of Human Dermal Fibroblasts under Diabetic Conditions through Reactive Oxygen Species Production via the PI3K/Akt-Rac1- JNK Pathways. *International journal of biological sciences* 11 (7): 845-859. doi: 10.7150/ijbs.11921.

Shingyochi, Y., Kanazawa, S., Tajima, S., Tanaka, R., Mizuno, H. and Tobita, M. (2017). A Low-Level Carbon Dioxide Laser Promotes Fibroblast Proliferation and Migration through Activation of Akt, ERK, and JNK. *PLOS ONE* 12 (1): e0168937. doi: 10.1371/journal.pone.0168937.

Silveira, P. C. L., Silva, L. A., Freitas, T. P., Latini, A. and Pinho, R. A. (2011). Effects of low-power laser irradiation (LPLI) at different wavelengths and doses on oxidative stress and fibrogenesis parameters in an animal model of wound healing. *Lasers in medical science* 26 (1): 125-131.

Skopin, M. D. and Molitor, S. C. (2009). Effects of near-infrared laser exposure in a cellular model of wound healing. *Photodermatology Photoimmunology & Photomedicine* 25 (2): 75-80.

Sperandio, F. F., Giudice, F. S., Corrêa, L., Pinto, D. S., Jr., Hamblin, M. R. and de Sousa, S. C. O. M. (2013). Low-level laser therapy can produce increased aggressiveness of dysplastic and oral cancer cell lines by modulation of Akt/mTOR signaling pathway. *Journal of biophotonics* 6 (10): 839-847. doi: 10.1002/jbio.201300015.

Srinivasan, S. and Avadhani, N. G. (2012). Cytochrome c oxidase dysfunction in oxidative stress. *Free Radical Biology and Medicine* 53 (6): 1252-1263.

Stahnke, T., Kowtharapu, B. S., Stachs, O., Schmitz, K.-P., Wurm, J., Wree, A., Guthoff, R. F. and Hovakimyan, M. (2017). Suppression of TGF-β pathway by pirfenidone decreases extracellular matrix deposition in ocular fibroblasts in vitro. *PLOS ONE* 12 (2): e0172592. doi: 10.1371/journal.pone.0172592.

Storz, P. and Toker, A. (2003). Protein kinase D mediates a stress-induced NF-κB activation and survival pathway. *The EMBO journal* 22 (1): 109-120.

Taflinski, L., Demir, E., Kauczok, J., Fuchs, P. C., Born, M., Suschek, C. V. and Opländer, C. (2014). Blue light inhibits transforming growth factor-β1-induced myofibroblast differentiation of human dermal fibroblasts. *Experimental Dermatology* 23 (4): 240-246. doi: 10.1111/exd.12353.

Tandon, A., Tovey, J. C., Sharma, A., Gupta, R. and Mohan, R. R. (2010). Role of transforming growth factor Beta in corneal function, biology and pathology. *Current molecular medicine* 10 (6): 565-578.

Tang, E., Khan, I., Andreana, S. and Arany, P. R. (2017). Laser-activated transforming growth factor-β1 induces human β-defensin 2: implications for laser therapies for periodontitis and peri-implantitis. *Journal of periodontal research* 52 (3): 360-367. doi: 10.1111/jre.12399.

Tasken, K. and Aandahl, E. M. (2004). Localized effects of cAMP mediated by distinct routes of protein kinase A. *Physiological reviews* 84 (1): 137-167.

Terblanche, U., Evans, D. H. and Abrahamse, H. (2009). Effect of Low Level Laser Therapy on human skin keratinocytes: peer reviewed short review. *Medical Technology SA* 23 (2): 23-29.

Tergaonkar, V. (2006). NFκB pathway: A good signaling paradigm and therapeutic target. *The international journal of biochemistry & cell biology* 38 (10): 1647-1653.

Tibbs, M. K. (1997). Wound healing following radiation therapy: a review. *Radiotherapy and Oncology* 42 (2): 99-106.

Tomic-Canic, M. (2005). Targeting the science within wounds. *Suppl Wounds* 26: 2-5.

Tredget, E. E., Wang, R., Shen, Q., Scott, P. G. and Ghahary, A. (2000). Transforming growth factor-beta mRNA and protein in hypertrophic scar tissues and fibroblasts: antagonism by IFN-alpha and IFN-gamma in vitro and in vivo. *Journal of Interferon & Cytokine Research* 20 (2): 143-152.

Tuner, J. and Hode, L. 2002. *Laser therapy: clinical practice and scientific background: a guide for research scientists, doctors, dentists, veterinarians and other interested parties within the medical field*: Prima Books AB.

Tyagi, S., Gupta, P., Saini, A. S., Kaushal, C. and Sharma, S. (2011). The peroxisome proliferator-activated receptor: A family of nuclear receptors role in various diseases. *Journal of advanced pharmaceutical technology & research* 2 (4): 236-240. doi: 10.4103/2231-4040.90879.

Usumez, A., Cengiz, B., Oztuzcu, S., Demir, T., Aras, M. H. and Gutknecht, N. (2014). Effects of laser irradiation at different wavelengths (660, 810,

980, and 1,064 nm) on mucositis in an animal model of wound healing. *Lasers in medical science* 29 (6): 1807-1813.

Vitor, L. L. R., Prado, M. T. O., Neto, N. L., de Oliveira, R. C., Santos, C. F., Machado, M. A. A. M. and Oliveira, T. M. (2018). Photobiomodulation changes type 1 collagen gene expression by pulp fibroblasts. *Laser Physics* 28 (6): 065603. doi: 10.1088/1555-6611/aabd16.

Vyas, B., Ishikawa, K., Duflo, S., Chen, X. and Thibeault, S. L. (2010). Inhibitory effects of hepatocyte growth factor and interleukin-6 on transforming growth factor-β1 mediated vocal fold fibroblast-myofibroblast differentiation. *Annals of Otology, Rhinology & Laryngology* 119 (5): 350-357.

Wall, I. B., Moseley, R., Baird, D. M., Kipling, D., Giles, P., Laffafian, I., Price, P. E., Thomas, D. W. and Stephens, P. (2008). Fibroblast dysfunction is a key factor in the non-healing of chronic venous leg ulcers. *Journal of Investigative Dermatology* 128 (10): 2526-2540.

Wang, I.-C., Chen, Y.-J., Hughes, D., Petrovic, V., Major, M. L., Park, H. J., Tan, Y., Ackerson, T. and Costa, R. H. (2005). Forkhead box M1 regulates the transcriptional network of genes essential for mitotic progression and genes encoding the SCF (Skp2-Cks1) ubiquitin ligase. *Molecular and cellular biology* 25 (24): 10875-10894.

Wang, L. and Jacques, S. L. (1992). Monte Carlo modeling of light transport in multi-layered tissues in standard C. *The University of Texas, MD Anderson Cancer Center, Houston*: 4-11.

Wang, X., Tian, F., Soni, S. S., Gonzalez-Lima, F. and Liu, H. (2016). Interplay between up-regulation of cytochrome-c-oxidase and hemoglobin oxygenation induced by near-infrared laser. *Scientific Reports* 6: 30540. doi: 10.1038/srep30540 https://www.nature.com/articles/srep30540#supplementary-information.

Weng, Y., Dang, Y., Ye, X., Liu, N., Zhang, Z. and Ren, Q. (2011). Investigation of irradiation by different nonablative lasers on primary cultured skin fibroblasts. *Clinical and Experimental Dermatology* 36 (6): 655-660. doi: 10.1111/j.1365-2230.2011.04043.x.

Werner, S. and Grose, R. (2003). Regulation of wound healing by growth factors and cytokines. *Physiological reviews* 83 (3): 835-870.

Willenborg, S., Ranjan, R., Krieg, T. and Eming, S. (2010). Chronic wounds and inflammation. *Adv Wound Care* 1: 259.

Witte, M. B. and Barbul, A. (2002). Role of nitric oxide in wound repair. *The American Journal of Surgery* 183 (4): 406-412.

Wong-Riley, M. T., Liang, H. L., Eells, J. T., Chance, B., Henry, M. M., Buchmann, E., Kane, M. and Whelan, H. T. (2005). Photobiomodulation directly benefits primary neurons functionally inactivated by toxins role of cytochrome c oxidase. *Journal of Biological Chemistry* 280 (6): 4761-4771.

Wong, V. W. and Crawford, J. D. (2013). Vasculogenic cytokines in wound healing. *BioMed Research International* 2013.

Wrana, J. L., Attisano, L., Wieser, R., Ventura, F. and Massagué, J. (1994). Mechanism of activation of the TGF-β receptor. *Nature* 370 (6488): 341.

Xiao, Z. S., Simpson, L. G. and Quarles, L. D. (2003). IRES-dependent translational control of Cbfa1/Runx2 expression. *Journal of cellular biochemistry* 88 (3): 493-505.

Xing, W., Guo, W., Zou, C.-H., Fu, T.-T., Li, X.-Y., Zhu, M., Qi, J.-H., Song, J., Dong, C.-H., Li, Z., Xiao, Y., Yuan, P.-S., Huang, H. and Xu, X. (2015). Acemannan accelerates cell proliferation and skin wound healing through AKT/mTOR signaling pathway. *Journal of Dermatological Science* 79. doi: 10.1016/j.jdermsci.2015.03.016.

Xue, L., Chiang, L., He, B., Zhao, Y.-Y. and Winoto, A. (2010). FoxM1, a forkhead transcription factor is a master cell cycle regulator for mouse mature T cells but not double positive thymocytes. *PloS one* 5 (2): e9229.

Yeh, M.-C., Chen, K.-K., Chiang, M.-H., Chen, C.-H., Chen, P.-H., Lee, H.-E. and Wang, Y.-H. (2017). Low-power laser irradiation inhibits arecoline-induced fibrosis: an in vitro study. *International journal of oral science* 9 (1): 38.

Yu, W., Naim, J. O., McGowan, M., Ippolito, K. and Lanzafame, R. J. (1997). Photomodulation of oxidative metabolism and electron chain enzymes in rat liver mitochondria. *Photochemistry and photobiology* 66 (6): 866-871.

Zagotta, W. N. and Siegelbaum, S. A. (1996). Structure and function of cyclic nucleotide-gated channels. *Annual review of neuroscience* 19 (1): 235-263.

Zein, R., Selting, W. and Hamblin, M. R. (2018). Review of light parameters and photobiomodulation efficacy: dive into complexity. *Journal of Biomedical Optics* 23 (12): 1-17, 17.

Zhang, H., Ackermann, A. M., Gusarova, G. A., Lowe, D., Feng, X., Kopsombut, U. G., Costa, R. H. and Gannon, M. (2006). The FoxM1 transcription factor is required to maintain pancreatic β-cell mass. *Molecular endocrinology* 20 (8): 1853-1866.

Zungu, I. L., Hawkins Evans, D. and Abrahamse, H. (2009). Mitochondrial responses of normal and injured human skin fibroblasts following low level laser irradiation--an in vitro study. *Photochem Photobiol.* 85 (4): 996. doi: 10.1111/j.1751-1097.2008.00523.x.

In: A Closer Look at Fibroblasts
Editor: Justin O'Shane

ISBN: 978-1-53616-977-5
© 2020 Nova Science Publishers, Inc.

Chapter 2

THE ROLE OF FIBROBLASTS IN OVARIAN CANCER

Rosekeila Simões Nomelini[*],
*Isa Beatriz Carminatti Batista,
Simone Paula Queiroz,
Ana Carolinne da Silva
and Eddie Fernando Candido Murta*

Research Institute of Oncology (IPON),
Department of Gynecology and Obstetrics,
Federal University of Triângulo Mineiro, Uberaba – MG, Brazil

ABSTRACT

Ovarian cancer corresponds to a heterogeneous group of malignancies, and the majority is of epithelial origin. Ovarian neoplasms are divided into several histological subtypes. A classification that takes into account the ovarian carcinogenesis model divides ovarian tumors into two groups: type I and type II neoplasias. Type I neoplasms include serous, clear cell,

---

[*] Corresponding Author's Email: rosekeila@terra.com.br; rosekeila.nomelini@pq.cnpq.br.

endometrioid, mucinous, cancers, and have milder growth characteristics with slow growth, and most often grow from an identifiable precursor. On the other hand, type II neoplasias are characterized by neoplasias that develop rapidly, are more aggressive and classified as high grade, emphasizing that high grade serous ovarian carcinoma is the most common type II neoplasia, characterizing almost 75% of all cancers of epithelial origin.

Fibroblasts are responsible for providing structural integrity to most tissues, producing the tissue's own basement membrane, providing a protective barrier around the epithelium, thus contributing to the polarity, functionality and specificity of the epithelium. Such factors indicate that fibroblasts are critical parts during wound healing and inflammation processes.

Several recent studies have demonstrated the importance of the tumor microenvironment in the processes of evolution and progression of ovarian cancer, and stroma plays an important role in the mechanisms of cancer progression. The most common cell types in the tumor microenvironment are cancer-associated fibroblasts (CAFs), which are often present at all stages of the disease's evolution. The role of fibroblasts in the progression of ovarian cancer is complex, but some functions have already been well understood, such as the ability to produce extracellular matrix, chemokines, cytokines, growth factors, and also stimulate angiogenic recruitment events of pericytes and endothelial cells. In this way, fibroblasts are decisive components in the evolution of cancer.

CAFs activation is directly related to inflammation and recruitment of inflammatory cells to the stroma. Two cancer-associated fibroblasts markers, alpha-smooth muscle actin and fibroblast activation protein alpha, may be found in the stromal microenvironment of benign and malignant ovarian epithelial neoplasms, and to relate their tissue expression with prognostic factors in ovarian cancer.

The objectives of this chapter are to demonstrate the importance of fibroblasts in the process of development and progression of ovarian cancer, helping to guide new studies that establish new therapeutic management targeted to the CAFs. In addition, we will address stromal fibroblasts as potential therapeutic targets and the effect of chemotherapy treatment on fibroblasts.

Keywords: fibroblasts, cancer-associated fibroblasts (CAFs), ovarian cancer, stroma

OVARIAN CANCER

Ovarian cancer is the fifth leading cause of gynecological cancer-related deaths in women worldwide. It includes a wide genetic and histological variety of tumors, including germinal, epithelial, and stromal sex cord tumors (Karnezis et al., 2017; Zhang et al., 2018).

One of the main features of ovarian cancer is its heterogeneity. Tumors of epithelial origin are the most frequent. Ovarian epithelial cancer can arise through the ovarian compartment and develop from flattened epithelial cells that cover the ovarian surface or form inclusion cysts (Kroeger and Drapkin 2017; Yang et al., 2017). These superficial epithelial cells modify and subdivide into four distinct histological types: 1. types resembling endometrial lining cells (endometrioids), 2. types resembling endocervical glands (mucinous), 3. types resembling fallopian tube (serous), and 4. types that originate from the vagina epithelium (clear cells) (Yang et al., 2017).

A more recent classification based on histological, molecular phenotypic and genotypic grades classifies ovarian neoplasms into two major groups. Low grade or type I ovarian neoplasms, which may be mucinous, endometrioid, serous, or clear cell histology. And high-grade or type II neoplasms, which may include endometrioid, serous or undifferentiated histological types (Yang et al., 2017). Elucidating the differentiation of these two large groups of ovarian neoplasms, type I neoplasms usually represent unilateral, large, cystic tumors. Type I neoplasms tend to behave indolently, with a good prognosis. They account for only 10% of ovarian cancer deaths and are often diagnosed at an early stage (I/II) (Kurman and Shih 2016; Yang et al., 2017). In contrast, type II ovarian neoplasms have substantially smaller volumes than type I neoplasms. They develop rapidly, are highly aggressive, and present at an advanced stage (III/IV) in more than 75% of cases. Another point to consider is the frequent occurrence of ascites episodes, which account for 90% of deaths from ovarian cancer (Kurman and Shih 2016; Yang et al., 2017).

Thus, analyzing the heterogeneity of ovarian cancer, further studies need to be performed in order to improve survival of ovarian epithelial cancer. And for this to occur, it is necessary to understand the molecular and pathological etiology of this cancer, thus being able to customize individual strategies for the treatment and optimization of early detection (Bast et al., 2009; Bowtell et al., 2015).

TUMOR MICROENVIRONMENT AND CANCER-ASSOCIATED FIBROBLASTS (CAFS)

The mechanism of cancer development encompasses a long and complex process, but some particularities have been well clarified, and these integrate six biological capacities acquired during the process of cancer development. Such capabilities include: sustaining proliferation signals, preventing growth suppressing agents, resisting cell death allowing replicative immortality, inducing angiogenesis, and activating metastasis and invasion processes (Hanahan and Weinberg 2011).

In addition to these capacities, neoplasias have other complex peculiarities, which occur through the recruitment of immune cells that acquire protumor characteristics. Similarly, there is the interaction of blood vessels, extracellular components and molecules that surround neoplastic cells. This context characterizes the "tumor microenvironment" or "stroma" (Hanahan and Weinberg 2011; Marsh et al., 2013).

In situ neoplastic lesions are limited within a basement membrane that divides epithelial cells from the underlying stroma. In the course of malignant evolution, cancer cells gain invasive features, break the basement membrane, and invade the stroma (Santi et al., 2018; Hanahan and Weinberg 2000). Fibroblasts originate from the primitive mesenchyme, their shape is elongated, and have active metabolism. They are characterized as the most abundant cell group in connective tissue, which degrade and produce extracellular matrix components. They also express fibronectins, elastins, laminins, collagens, metalloproteinases, integrins, and a variety of other extracellular matrix proteins, which are expressed in a tissue-specific manner (Marsh et al., 2013).

As a result, fibroblasts are responsible for providing structural integrity to most tissues, producing the tissue's own basement membrane, providing a protective barrier around the epithelium, thus contributing to the polarity, functionality and specificity of the epithelium. Such factors indicate that fibroblasts are critical parts during wound healing and inflammation processes (Marsh et al., 2013).

In normal tissues, the stroma is filled by few fibroblasts that are within the extracellular matrix (Marsh et al., 2013). In contrast, *in situ* tumors, the tumor and stromal cells can communicate across the basement membrane, leading to an evolution in tumor development and changes in the microenvironment (Hanahan and Weinberg 2000; Hanahan and Weinberg 2011). In this tumor evolutionary process, the stroma presents a high number of fibroblasts, which are now called 'cancer-associated fibroblasts' (CAFs) (Santi et al., 2018).

CAFs are a heterogeneous cell group, that play key roles in building a tumor-modulating microenvironment of surrounding cell behavior and functions during the progression of the tumor process. For this reason, CAFs have been named "architects of the pathogenesis of cancer" (Santi et al., 2018).

In addition, by analyzing all functionalities performed by CAFs, they have been recognized as essential regulators of tumor progression processes and induce response to treatments. A better understanding of the biology of CAFs is needed to bring concrete strategies for relating this cell group to carcinogenesis processes (Santi et al., 2018).

CANCER-ASSOCIATED FIBROBLASTS AND OVARIAN CANCER

Little is known about the origin of CAF, but it is believed to originate from ovarian cell stroma. The ovary has some unique characteristics that may be useful for understanding the origin of the CAFs. Inflammation and endometriosis may alter the origin of CAFs. Other hypotheses have been formulated to explain the origin of CAF in the ovarian tumor, such as: a) the ovary is an isolated organ in the abdominal cavity, making it difficult for other neighboring cells to invade it, and is related to the peritoneum and supported by a discrete mesovary; b) It is a small organ that can increase in volume when the tumor is present, indicating that practically the tumor stroma is newly formed (Fujisawa et al., 2018); c) ovarian stromal cells have a specific protein, FOXL2 (Forkhead box protein L2), important for ovarian development and function, can be used as a marker; d) the fibroblast markers such as smooth muscle alpha actin and fibroblast activating protein were also found in ovarian tumor cells that were positive for FOXL-2 (Fujisawa et al., 2018).

In addition, FOXL2 acts as a precursor to other expressions in tumor stroma. In other words, by analyzing the interaction (FOXL2)/tumor stroma by immunohistochemistry, the expression of FOXL2 in ovarian lesions was evident, as well as two stroma markers were expressed: alpha-smooth muscle actin and fibroblast activation protein alpha (Fujisawa et al., 2018). Other studies have demonstrated FOXL2 expression in a restricted manner only in stromal cells, and some other cell types in the female genital tract (Governini et al., 2015; Bellessort et al., 2015; Eozenou et al., 2012). Thus, it was possible to observe the interaction of ovarian stromal FOXL2 protein with CAFs, indicating that FOXL2 can be considered a specific marker of ovarian stromal cells, helping to a possible understanding of the origin of CAFs (Fujisawa et al. 2018).

A study evaluated these two cancer-associated fibroblasts markers, alpha-smooth muscle actin and fibroblast activation protein alpha, in the stromal microenvironment of benign and malignant ovarian epithelial neoplasms, and to relate their tissue expression with prognostic factors in ovarian cancer. Fibroblast activation protein alpha immunostaining was more intense in malignant neoplasms than in benign ovarian neoplasms, as well as in moderately differentiated and undifferentiated ovarian carcinomas compared to well differentiated neoplasms, suggesting to be a marker of worse prognosis (da Silva et al., 2018).

Finally, in addition to all the contribution of CAFs in the ovarian cancer development and progression processes, the role of CAFs in drug resistance processes in ovarian cancer cells is also highlighted. A study by Deying et al. (2017) demonstrated that supernatants were evaluated by isolating CAFs in tumor tissues from ovarian cancer patients. Hepatocyte growth factor (HGF) was observed to be highly expressed in CAF supernatants, promoting cell proliferation as well as drug resistance in ovarian cancer. These findings suggest that CAFs present in the tumor microenvironment were essential in drug resistance in ovarian cancer cells.

INTERACTION OF FIBROBLASTS IN TUMOR STROMA

Malignant tumors, regardless of the organ of origin, have two basic constituents: the parenchyma, made up of neoplastic cells, and the stroma (representing 50% of the tumor), composed of non-tumorigenic cells such as fibroblasts, macrophages, blood and lymphatic vessels, nerves, inflammatory cells and extracellular matrix (ECM). The importance of the tumor microenvironment's role in tumor development and growth is increasingly evident, as the interaction between the different neoplastic cell types and the surrounding tissue (stroma) plays an important role in regulating epithelial cell behavior and not just supporting structural function (Joyce 2005).

Although its origin and function remain controversial, it is known that the tumor stroma is capable of providing the main characteristics and behaviors of cancer, including the escape of immune surveillance, angiogenesis, invasion, metastasis and response to chemotherapy. There is evidence that the higher the stromal content, referred to as the "high tumor stroma proportion", and the more active it is, the worse the prognosis and greater resistance to chemotherapy (Lou et al., 2019).

In some types of ovarian tumor, the stroma is called a "functioning stroma" due to the production of sex hormones and endocrine function of luteinization and hyperthecosis (Narikiyo et al., 2018). An association between mucinous tumors and "functioning stroma" was found in one study

(Kato et al., 2013; Katoh et al., 2012), reported that this stroma is more common in endometrioid carcinomas, resembling the stroma of sexual cord tumor.

The main component of the tumor stroma is fibroblast, known as "cancer-associated fibroblast - CAF", which creates a favorable microenvironment for extracellular matrix invasion, metastasis and remodeling. Its activation is directly related to inflammation and recruitment of inflammatory cells to the stroma. It remains unknown whether fibroblasts from the primary tumor site migrate with the tumor cells to other organs where they will be metastasized or these cells migrate alone and recruit CAF into the organ to be metastasized (Fujisawa et al., 2018). The histopathological study showed that fibroblasts could only metastasize via direct implantation in the peritoneal cavity, and could neither disseminate hematogenous nor lymphatic pathways. Other studies have indicated that CAFs can spread to other organs by altering the presence or absence of FOXL-2.

In addition to FOXL-2 protein, other molecules are also involved in the origin and action of CAF. Dicer is a nuclease enzyme capable of cleaving RNA into smaller portions, giving rise to microRNAs, playing an important role in cell regulatory function, maturation and growth. Such a molecule is a tumor suppressor in the epithelial ovarian tumor, and its expression correlates positively with patient survival for various cancers. It has been paradoxically seen that the enzyme in large quantities maintains tumor stroma reactivity and inflammation. In fibroblasts, their presence contributes to activation and infiltration of inflammation through the transcription factor NFkB. Thus, it is suggested that the axis between Dicer and NFkB may be a target for treatment (Yang et al., 2017).

Thus, it is evident the undefined origin of fibroblasts and their action on the tumor stroma, but it is agreed that this favors the invasion and metastasis of cancer. Several molecules present in fibroblasts have been studied to become a possible treatment target, but still require further studies for elucidation.

STROMAL FIBROBLASTS AS POTENTIAL THERAPEUTIC TARGETS

It is believed that there are many subtypes of fibroblasts deposited in the extracellular matrix and therefore the difficulty of obtaining a good response to treatment when they are targeted. CAFs paralyze regulatory CD4, CD8, natural killers, T cells and secrete CXCL12 (C-X-C motif chemokine ligand 1) to keep T cells containing the CXCR4 (C-X-C chemokine receptor type 4) away from the tumor region (Zhang et al., 2019). CXCR4 has a single ligand, CXCL12, and is expressed on most hematopoietic cells, including neutrophils, monocytes, T lymphocytes and B precursors, bone marrow, dendritic cells, Langerhans cells, macrophages, vascular endothelial cells, and neurons of the central and peripheral nervous system. It is a receptor implicated in platelet formation and chemotaxis (Guerreiro et al., 2011). In addition, CAFs secrete CCL5, CCL2 and CCL17 (C-C motif chemokines ligands 5, 2 and 17) to recruit monocytes and regulatory T cells, resulting in immunosuppression.

Normal ovarian epithelial cells can be modified by tumor cells, inducing them to produce growth factors, chemokines and enzymes that degrade the cell matrix, allowing tumor proliferation and invasion. These normal cells may further assist the tumor in resisting conventional treatments. In addition, changes in the stromal microenvironment, such as increased interstitial fluid pressure and variations in vascular flow, may decrease drug arrival at the site, thereby promoting chemoresistance (Joyce 2005). Due to such resistance described, researchers have been looking for new ways to control tumor progression by diverting tumor cells.

The transforming growth factor beta (TGF-β) is known to act on development, proliferation and apoptosis by activating fibroblasts. The stromal microenvironment is considered to have more genetically stable cells, as opposed to neoplastic cells, which can often accumulate mutations and thus produce tolerance to anti-cancer drugs (Joyce 2005). Thus, it could be a good therapeutic target by inhibiting (TGF-β) signaling, suppressing the interaction between stroma and tumor cell in ovarian mesenchymal neoplasms, reducing tumor growth and metastasis. (Zhang et al., 2018). However, proper stromal cell control must be found, as there is a fine line between their inhibitory and tumor advancement functions.

The SNAI2 (snail family transcriptional repressor 2) protein is also expressed in stromal fibroblasts and was correlated with stromal activation and worse prognosis, as this molecule transforms normal fibroblasts into CAFs. Such molecule has its levels increased and accumulated in fibroblast nuclei after exposure to ECM, exerting continuous activation on them (Yang et al., 2017).

A molecular profile of fibroblasts from a normal ovary was compared to a high tumor grade ovarian of the serous type, in which about 2,300 different CAF genes were identified. One of the main identified was the TGF-β-regulated connective tissue growth factor (CTGF) secreted in the ovarian tumor stroma and also in the ovarian epithelium with migration and metastasis through the peritoneal adhesion of ovarian cancer cells. During the research, it was found that its expression was undetectable in the stroma and epithelium of normal ovaries. Therefore, the higher its levels, the worse the tumor prognosis. Firstly in pancreatic cancer, monoclonal antibody FG-3019 has been studied as a possible therapy to block the effects of CTGF and studies have shown that the antibody reduces tumor cell migration and adhesion, at least *in vitro* (Jones et al., 2015).

Another study combined the evaluation of CTGF together with Cyr61 (cysteine-rich growth factor 61), whose expression occurs in endothelial cells and fibroblasts and noted the presence of Cyr61 in benign ovarian stromal fibroblasts and epithelium of the ovarian tubes; on the contrary, its expression was not seen in the epithelium of this organ. This protein has been found in serous but not endometrioid and clear cell tumors, which aids in the investigation of new therapeutic targets specific to each ovarian cancer subtype, as Cyr61 is associated with a chemoresistance phenotype. A study by Bartel et al. (2012), demonstrated that the expression of such a protein is estrogen dependent in both ovaries and breast and its onset, as well as CTGF, is reduced in hypoxia situations (Bartel et al., 2012).

SPARC or osteonectin (secreted protein acidic and rich in cysteine) is a glycoprotein that has different expressions depending on normal or neoplastic tissue. Although its real contribution to tumors remains unknown, endothelial cells promote increased vascularization, alteration of cytokine activity, and stimulate the secretion of metalloproteases that will cause tissue transformations. In melanomas, it was seen that SPARC acts stimulatory and when inhibited, there is a reduction in invasive potential. In contrast, in the study, it was observed that in ovarian tumors it exerts inhibitory effect by acting as a tumor suppressor. It was also observed that SPARC mRNA expression was not detected in tumor ovarian cells or benign ovarian tissues; however, elevated levels were found in stromal cells and metastatic lymph node interface. This may indicate that SPARC activation present in the stroma occurs and even depends on the metastasis site and that the total level of this glycoprotein depends on the amount of stroma present (Brown et al., 1999).

The relationship between SPARC and platelet-derived growth factor (PDGF) was also studied, the latter being known to interfere with angiogenesis and lymphatic vessel development in various types of tumors. Skobe and Fuseng (1998) observed that PDGF-activated stromal cells would support cell growth and transformation of benign into tumorigenic cells. SPARC glycoprotein binds to and inhibits PDGF, which may infer that its expression is a protective response to platelet growth factor actions (Brown et al., 1999).

Another molecule studied was fibroblast growth factor 4 receptor (FGFR4), which is one of the receptors for fibroblast growth factor 1 (FGF1, which is considered a gene for ovarian tumor progression). Since FGF1 and FGFR4 overexpression correlates with decreased survival, a further change in FGFR4 expression could cause activation in the FGF receptor axis, impacting tumor progression, especially in serous ovarian tumors. Zaid et al. (2012) realized that targeting FGFR4 in mice was able to slow serous tumor growth after inhibiting binding to such receptor (Zaid et al., 2013).

THE EFFECT OF CHEMOTHERAPY TREATMENT ON FIBROBLASTS

Chemotherapy treatment acts not only on neoplastic cells but also on normal somatic cells. Platinum and taxane analogs used in the treatment of ovarian cancer are known to generate various side effects and local and systemic complications. It is also known that drug resistance leading to an insufficient response to therapy is common in patients with ovarian cancer and usually leads to recurrence of the disease (Luvero et al., 2014). Research has already shown the association between the complications of chemotherapy treatment and the deleterious effects of chemotherapy on non-cancerous cells and tissues (Lui et al., 2015).

In this context, it is important to highlight the effect of chemotherapy on healthy fibroblasts. The first changes are abnormalities in general metabolism, especially in cellular energy. In a study that evaluated the response of normal stromal fibroblasts submitted to 12 common chemotherapeutic agents, including cisplatin, carboplatin and paclitaxel, showed that cells maintained under this regimen consume more glucose and produce more lactate, leading to acidification of the extracellular environment (Peiris-Pages et al., 2015).

Another feature of stromal fibroblasts submitted to cisplatin, carboplatin and paclitaxel is an induction of their phenotypic transformation into cancer-associated fibroblasts (CAFs) (Peiris-Pages et al., 2015). Evidence suggests that CAFs differ from normal fibroblasts in their molecular characteristics and ability to withstand the elements of cancer progression (Orimo et al., 2006).

The presence and activity of CAFs may play a role in cancer cell resistance to cisplatin. Cisplatin-submitted fibroblasts exhibit a regulated production of IL-11, which causes decreased susceptibility of cancer cells to cisplatin-induced apoptosis (Tao et al., 2016). Another resistance pathway involves the ability of fibroblasts to restrict cisplatin nuclear accumulation in cancer cells, which is associated with increased levels of glutathione and cysteine secreted by these cells in the environment (Wang et al., 2016).

In addition to the contribution of drug-induced CAFs to chemoresistance, cells with myofibroblastic characteristics are also known to promote tissue fibrosis (Klingberg et al., 2013). Considering that chemotherapy for ovarian cancer usually follows cytoreductive surgery - which increases the risk of developing intraperitoneal adhesions and fibrosis - the profibrotic effects generated by normal peritoneal fibroblasts due to their exposure to paclitaxel may compromise proper healing of the abdomen and indicate a increased risk of tissue fibrosis in the future (Sato et al., 2014). However, the effect of paclitaxel on profibrotic potential appears to be a cell-specific response, possibly limited to fibroblasts (Mikula-Pietrasik et al., 2016).

CONCLUSION

Ovarian cancer is one of the most common causes of gynecological cancer-related deaths in women worldwide. Thus, a better understanding of the mechanisms of development and progression of malignant ovarian neoplasia is necessary in the establishment of new therapeutic strategies. Activation of the stroma microenvironment has been shown to be an important factor in cancer progression. Stromal cells can control tumor growth and invasiveness, are related to the immune response and play an important role in the behavior of various types of neoplasia. CAFs are a heterogeneous cell group that play key roles in the development of a tumor microenvironment and are considered in several studies as indicators of worse prognosis in ovarian cancer. They are considered an active subpopulation of stromal component fibroblasts, which contribute to tumor cell proliferation and angiogenesis. Thus, further studies related to these fibroblasts may lead in the future to the discovery of new potential target therapies for this disease.

REFERENCES

Bartel, F., Balschun, K., Gradhand, E., Strauss, H. G., Dittmer, J. & Hauptmann, S. (2012). Inverse expression of cystein-rich 61 (Cyr61/CCN1) and connective tissue growth factor (CTGF/CCN2) in borderline tumors and carcinomas of the ovary. *International journal of gynecological pathology.*, *31*, 405–415.

Bast, R. C., Jr. Hennessy, B. & Mills, G. B. (2009). The biology of ovarian cancer: new opportunities for translation. *Nature Reviews Cancer*, *9*, 415-28.

Bellessort, B., Bachelot, A., Heude, É., Alfama, G., Fontaine, A., Le Cardinal, M., Treier, M. & Levi, G. (2015). Role of Foxl2 in uterine maturation and function. Hum *Human molecular genetics.*, Jun 1, *24*(11), 3092-103.

Bowtell, D. D., Böhm, S., Ahmed, A. A., Aspuria, P. J., Bast, R. C., Jr. Beral, V., Berek, J. S., Birrer, M. J., Blagden, S., Bookman, M. A., Brenton, J. D., Chiappinelli, K. B., Martins, F. C., Coukos, G., Drapkin, R., Edmondson, R., Fotopoulou, C., Gabra, H., Galon, J., Gourley, C., Heong, V., Huntsman, D. G., Iwanicki, M., Karlan, B. Y., Kaye, A., Lengyel, E., Levine, D. A., Lu, K. H., McNeish, I. A., Menon, U., Narod, S. A., Nelson, B. H., Nephew, K. P., Pharoah, P., Powell, D. J., Jr. Ramos, P., Romero, I. L., Scott, C. L., Sood, A. K., Stronach, E. A. & Balkwill, F. R. (2015). Rethinking ovarian cancer II: reducing mortality from high-grade serous ovarian cancer. *Nature Reviews Cancer*, Nov, *15*(11), 668-79.

Brown, T. J., Shaw, P. A., Karp, X., Huynh, M. H., Begley, H. & Ringuette, M. J. (1999). Activation of SPARC expression in reactive stroma associated with human epithelial ovarian cancer. *Gynecologic oncology.*, *75*, 25–33.

da Silva, A. C., Jammal, M. P., Etchebehere, R. M., Murta, E. F. C. & Nomelini, R. S. (2018). Role of Alpha-Smooth Muscle Actin and Fibroblast Activation Protein Alpha in Ovarian Neoplasms. *Gynecologic and obstetric investigation*, *83*(4), 381-387.

Deying, W., Feng, G., Shumei, L., Hui, Z., Ming, L. & Hongqing, W. (2017). CAF- derived HGF promotes cell proliferation and drug resistance by up-regulating the c-Met/PI3K/Akt and GRP78 signalling in ovarian cancer cells. *Bioscience reports.* Ap 10, *37*(2).

Eozenou, C., Vitorino Carvalho, A., Forde, N., Giraud-Delville, C., Gall, L., Lonergan, Fujisawa, M., Moh-Moh-Aung, A., Zeng, Z., Yoshimura, T., Wani, Y. & Matsukawa, A. (2018). Ovarian stromal cells as a source of cancer-associated fibroblasts in human epithelial ovarian cancer: A histopathological study. *Public Library of Science,* Oct 10, *13*(10), e0205494.

Governini, L., Carrarelli, P., Rocha, A. L., Leo, V. D., Luddi, A., Arcuri, F., Piomboni, P., Chapron, C., Bilezikjian, L. M. & Petraglia, F. (2014). FOXL2 in human endometrium: hyperexpressed in endometriosis. *Reproductive sciences,* Oct, *21* (10), 1249-55.

Guerreiro, R., Santos-Costa, Q. & Azevedo-Pereira, J. M. (2011). The chemokines and their receptors: characteristics and physiological functions. *Acta médica portuguesa,* Dec 24, *4,* 967-76.

Hanahan, D. & Weinberg, R. A. (2000). The hallmarks of cancer. *Cell,* Jan 7, *100*(1), 57-70.

Hanahan, D. & Weinberg, R. A. (2011). Hallmarks of cancer: the next generation. *Cell,* Mar 4, *144*(5), 646-74.

Jones, K. M., Gloss, B. S., Murali, R., Chang, D. K., Colvin, E. K., Jones, M. D., Yuen, S., Howell, V. M., Brown, L. M., Wong, C. W., Spong, S. M., Scarlett, C. J., Hacker, N. F., Shosh, S., Mok, S. C., Birrer, M. J. & Samimii, G. (2015). Connective tissue growth factor as a novel therapeutic target in high grade serous ovarian cancer. *Oncotarget,* Dec 29, *6*(42), 44551-62.

Joyce, J. A. (2005). Therapeutic targeting of the tumor microenvironment. *Cancer Cell, 7,* 513–520.

Karnezis, A. N., Cho, K. R., Gilks, C. B., Pearce, C. L. & Huntsman, D. G. (2017). The disparate origins of ovarian cancers: pathogenesis and prevention strategies. *Nature Reviews Cancer,* Jan, *17*(1), 65-74.

Kato, N., Hayasaka, T., Takeda, J., Osakabe, M. & Kurachi, H. (2013). Ovarian tumors with functioning stroma: A clinicopathologic study with special reference to serum estrogen level, stromal morphology and aromatase expression. *International journal of gynecological pathology, 32*, 556–561.

Katoh, T., Yasuda, M., Hasegawa, K., Kozawa, E., Maniwa, J. & Sasano, H. (2012). Estrogen-producing endometrioid adenocarcinoma resembling sex cord-stromal tumor of the ovary: A review of four postmenopausal cases. *Diagnostic pathology.*, *7*, 164.

Klingberg, F., Hinz, B. & White, E. S. (2013). The myofibroblast matrix: implications for tissue repair and fibrosis. *The Journal of pathology.*, Jan, *229*(2), 298-309.

Kroeger, P. T. Jr. & Drapkin, R. (2017). Pathogenesis and heterogeneity of ovarian cancer. *Current Opinion in Obstetrics and Gynecology,* Feb, *29*(1), 26-34.

Kurman, R. J. & Shih, IeM. (2016). The Dualistic Model of Ovarian Carcinogenesis: Revisited, Revised, and Expanded. *The American journal of pathology, 186*, 733-47.

Liu, B., Ezeogu, L., Zellmer, L., Yu, B., Xu, N. & Joshua Liao, D. (2015). Protecting the normal in order to better kill the cancer. *Cancer medicine.*, *4*(9), 1394-1403.

Lou, E., Vogel, R. I., Hoostal, S., Klein, M., Linden, M. A., Teoh, D. & Geller, M. A. (2019). Tumor-Stroma Proportion as a Predictive Biomarker of Resistance to Platinum-Based Chemotherapy in Patients with Ovarian Cancer. *JAMA oncology*, Jun 1.

Luvero, D., Milani, A. & Ledermann, J. A. (2014). Treatment options in recurrent ovarian cancer: latest evidence and clinical potential. *Therapeutic advances in medical oncology, 6*(5), 229–239.

Marsh, T., Pietras, K. & McAllister, S. S. (2013). Fibroblasts as architects of cancer pathogenesis. *Biochimica et biophysica acta. General subjects,* Jul, *1832*(7), 1070-8.

Mikula-Pietrasik, J., Uruski, P., Sosinska, P., Maksin, K., Piotrowska-Kempisty, H., Kucinska, M., Murias, M., Szubert, S., Wozniak, A., Szpurek, D., Sajdak, S., Piwocka, K., Tykarski, A. & Ksiazek, K. (2016). Senescent peritoneal mesothelium creates a niche for ovarian cancer metastases. *Cell death & disease*, Dec 29, *7*(12), e2565.

Narikiyo, M., Yano, M., Kamada, K., Katoh, T., Ito, K., Shuto, M., Kaano, H. & Yasuda, M. (2019). Molecular association of functioning stroma with carcinoma cells in the ovary: A preliminary study. *Oncology letters, 17*(3), 3562-3568.

Orimo, A. & Weinberg, R. A. (2006). Stromal fibroblasts in cancer: a novel tumor-promoting cell type. *Cell Cycle, 5*(15), 1597-1601.

Auguste, P., Charpigny, A., Richard, G., Pannetier, C. M. & Sandra, O. (2012). FOXL2 is regulated during the bovine estrous cycle and its expression in the endometrium is independent of conceptus-derived interferon tau. *Biology of reproduction.*, Aug 9, *87*(2), 32.

Peiris-Pages, M., Sotgia, F. & Lisanti, M. P. (2015). Chemotherapy induces the cancer-associated fibroblast phenotype, activating paracrine Hedgehog-GLI signalling in breast cancer cells. *Oncotarget, 6*(13), 10728–10745.

Santi, A., Kugeratski, F. G. & Zanivan, S. (2018). Cancer Associated Fibroblasts: The Architects of Stroma Remodeling. *Proteomics*, Mar. *18*(5-6), e1700167.

Sato, S. & Itamochi, H. (2014). Neoadjuvant chemotherapy in advanced ovarian cancer: latest results and place in therapy. *Therapeutic advances in medical oncology*, Nov, *6*(6), 293-304.

Skobe, M. & Fuseng, N. E. (1998). Tumorigenic conversion of imortal human keratinocytes through stromal cell activation. *Proceedings of the National Academy of Sciences of the United States of America*, Feb 3, *95*(3), 1050-5.

Tao, L., Huang, G., Wang, R., Pan, Y., He, Z., Chu, X., Song, H. & Chen, L. (2016). Cancer-associated fibroblasts treated with cisplatin facilitates chemoresistance of lung adenocarcinoma through IL-11/IL-11R/STAT3 signaling pathway. *Scientific reports*, Dec 6, *6*, 38408.

Wang, W., Kryczek, I., Dostal, L., Lin, H., Tan, L., Zhao, L., Lu, F., Wei, S., Maj, T., Peng, D., He, G., Vatan, L., Szeliga, W., Kuick, R., Kotarski, J., Tarkowski, R., Dou, Y., Rattan, R., Munkarah, A., Liu, J. R. & Zou, W. (2016). Effector T cells abrogate stroma-mediated chemoresistance in ovarian cancer. *Cell, 165*(5), 1092-1105.

Yang, W. L., Lu, Z. & Bast, R. C. Jr. (2017). The role of biomarkers in the management of epithelial ovarian cancer. *Expert Review of Molecular Diagnostics*, Jun, *17*(6), 577-591.

Yang, Z., Jin, P., Xu, S., Zhang, T., Yang, X., Li, X., Wei, X., Sun, C., Chen, G., Ma, D. & Gao, Q. (2017). Dicer reprograms stromal fibroblasts to a pro-inflamatory and tumor promoting phenotype in ovarian cancer. *Cancer Letters*, (415), 20-29.

Yang, Z., Yang, X., Xu, S., Jin, P., Li, X., Wei, X., Liu, D., Huang, K., Long, S., Wang, Y., Sun, C., Chen, G., Hu, J., Meng, L., Ma, D. & Gao, Q. (2017). Reprogramming of stromal fibroblasts by SNAI2 contributes to tumor desmoplasia and ovarian cancer progression. *Molecular cancer.*, *16*, 163.

Zaid, T. M., Yeung, T. L., Thompson, M. S., Leung, C. S., Harding, T., Co, N. N., Schmandt, R. S., Kwan, S. Y., Rodriguez-Aguay, C., Lopez-Berestein, G., Sood, A. K., Wong, K. K., Birrer, M. J. & Mok, S. C. (2013). Identification of FGFR4 as a potential therapeutic target for advanced-stage, high-grade serous ovarian cancer. *Clinical cancer research, 19*, 809–820.

Zhang, B., Chen, F., Xu, Q., Han, L., Xu, J., Gao, L., Sun, X., Li, Y., Li, Y., Qian, M. & Sun, Y. (2018). Revisiting ovarian cancer microenvironment: a friend or a foe? *Protein & Cell.*, Aug, *9*(8), 674-692.

Zhang, Q., Wang, C. & Cliby, W. A. (2019). Cancer-associated stroma significantly contributes to the mesenchymal subtype signature of serous ovarian cancer. *Gynecologic oncology*, Feb, *152*(2), 368-374.

Zhang, Y. F., Jiang, S. H., Hu, L. P., Huang, P. Q., Wang, X., Li, J., Zhang, X. L., Nie, H. Z. & Zhang, Z. G. (2019). Targeting the tumor microenvironment for pancreatic ductal adenocarcinoma therapy. *Chinese clinical oncology*, *8*(2), 18.

In: A Closer Look at Fibroblasts
Editor: Justin O'Shane

ISBN: 978-1-53616-977-5
© 2020 Nova Science Publishers, Inc.

Chapter 3

ANALYSIS OF THE EFFECT OF AN ANDIROBA, COPAÍBA AND GUARANÁ COMBINATION ON *IN VITRO* AND *IN VIVO* SCAR MODELS

Euler Esteves Ribeiro Filho[1], Bárbara Osmarin Turra[2], Bruna Chitolina[2], Beatriz Sadigursky Cunha[2], Cibele Ferreira Teixeira[2], Verônica Farina Azzolin[3], Ednea Aguiar Maia Ribeiro[3], Euler Esteves Ribeiro[2], Raquel de Souza Praia[3], Juscimar Carneiro Nunes[1], Ivana Beatrice Mânica da Cruz[2] and Fernanda Barbisan[2],*

[1]Universidade Federal do Amazonas, Manaus, Amazonas, Brazil
[2]Universidade Federal de Santa Maria, Santa Maria, Rio Grande do Sul, Brazil
[3]Fundação Universidade Aberta da Terceira Idade do Amazonas, Manaus- Amazonas, Brazil

---

* Corresponding Author's Email: fernandabarbisan@gmail.com.

ABSTRACT

Introduction: The skin is the largest organ of the human body and has the functions of coating, protecting and interacting with the external environment. Thus, the healing of the skin is a fundamental element for human health. The effectiveness of the healing process can further be enhanced by the development of products originating from plants with traditional use. Thus, in this study we searched for a healing compound based on the oils of andiroba (*Carapa guianensis*), copaíba (*Copaifera langsdorffii*) and guaraná (*Paullinia Cupana*), three plants of Amazonian origin that have potential healing effect through modulation of the inflammatory phase and proliferation of fibroblasts, with a consequent pro-cicatricial potential.

Objective: To evaluate the effect of the combination of andiroba, copaíba and guarana (ACG®), in the form of biphasic oil and emulsion, through the analysis of *in vitro* wound healing models (using dermal fibroblasts) and *in vivo* (with earthworms *Eisenia fetida* as a model of tissue regeneration via surgical removal of the caudal region).

Methodology: The antioxidant and genotoxic/genoprotective capacity of ACG® were evaluated in non-cellular tests. The line of dermal fibroblasts HFF-1 was cultured under standard conditions, and the stratch assay (*in vitro* model of skin lesion) was performed in the cells, that then were treated with ACG® in concentration of $2\mu L/mL$ and after 24/72 hours of incubation, and the cell migration and the proliferative, oxidative and inflammatory markers were analysed. In the *in vivo* model with Californian earthworms (*Eisenia fetida*), the animals had a caudal removal and were immediately treated with ACG®, in the form of biphasic oil and emulsion, and after 7 days the analysis of regeneration patterns was performed. Statistical analysis was performed by one-way analysis of variance followed by Tukey's post hoc. Tests with $p < 0.05$ were considered significant.

Results: ACG® showed strong antioxidant activity, and was not genotoxic. In front of the cellular model, the product presented proliferative, antioxidant and anti-inflammatory action. In the *in vivo* tests, the biphasic oil was effective in the renegeration process.

Conclusion: Despite the methodological limitations, we consider the data obtained to be relevant, once it was tested in the form of ACG® for the first time the combination of andiroba, copaíba and guaraná in cicatricial processes. The results show antioxidant, anti-inflammatory and pro-healing effects of the compound developed. Further studies must be performed to prove these results; however, we believe that in the future the development of a commercial product with clinical purposes may be a reality.

The results that will be presented come from a master's dissertation, presented at the Universidade Federal do Amazonas.

INTRODUCTION

Skin wound healing is a dynamic and complex process, coordinated by a cascade of cellular, molecular, and biochemical events that interact and determine tissue reconstitution. The orderly sequence of scar events involves four steps: (1) hemostasis; (2) inflammation; (3) proliferation; and (4) remodeling (Broughton et al. 2006; Ho et al. 2017).

The bed of an open wound needs to be filled, and this process takes place from two different and complementary mechanisms. The first involves the anatomical nature of the wound, which will provide stimuli that lead to the migration and proliferation of cells (fibroblasts, keratinocytes) to the center of the lesion from the uninjured tissue margins. This phenomenon is called "free neighborhood effects." The second involves the movement of the wound's margins toward each other as if under an invisible tractive force. This phenomenon occurs due to the differentiation of some fibroblasts from the wound margins to contractile-capable fibroblasts (myofibroblasts) (Balbino et al. 2005).

It is important to note that the wound surface, as it is oxygenated and well moistened, accelerates the process of epithelial cell migration to the center of the wound. When epithelial cells meet the crust of the wound, their migration velocity is delayed, a mechanism known as "contact inhibition" is triggered, and the cells return to their original cytological pattern. At the end of this phase, the wound bed is already completely filled with granulation tissue, the circulation has been reestablished by angiogenesis and the lymphatic network is also beginning to regenerate. Slowly, the granulation tissue is enriched with a greater amount of collagen fibers, which provides the appearance of the scar of the injured region as a result of the accumulation of fibrous mass at the site (Balbino et al. 2005).

It is important to stress that repair is the closure of scar-forming wounds, and regeneration is the perfect closure of damaged, scar-free tissue.

Potential Use of Amazonian Extracts in Healing

The use of medicinal plants to treat aspects related to human health and aesthetics has been reported for millennia in traditional medicine around the world. Currently, it is estimated that 25% of all traditional drugs are derived directly or indirectly from medicinal plants. For certain classes of pharmaceuticals, such as antitumor and antimicrobial drugs, this percentage may be greater than 60% (Brazil 2012).

A significant number of plant species represents a source of molecules with wide potential in pharmacological applications. These medicinal plants are important for pharmacological research and the development of new drugs, not only when their constituents are used directly as therapeutic agents, but also as raw materials for synthesis or models for pharmacologically active compounds (Brazil 2016).

However, it is important that research is carried out in all its phases so that the use of healing plants can be based on the assumptions of evidence-based medicine. Therefore, researching the potential of herbal drugs is an emergency and a challenging responsibility that is indispensable in the development of new technologies related to skin lesion care (Amirkia and Heinrich 2015; Mabona and Van Vuuren 2013).

In surgical procedures, injured cells are replaced by scar tissue, which is made up of collagen fibers. This process may eventually occur in an unorganized manner, producing ripples, depressions, thightning associated with major postoperative complications, including hematoma formation, seroma, delayed cut healing, loss of nipple sensitivity, ischemia or necrosis, rehypertrophy between others (Barr et al. 2016; Hammond and Kim 2016; Misani and De Mey 2016; Shestak and Davidson 2016). Therapeutic strategies to minimize these complications have been investigated, especially those that increase hyaluronic acid production, hydrate and are anti-inflammatory (Mahedia et al. 2016; Park et al. 2014; Wen et al. 2010).

There are many plant extracts that have potential healing effect and are used in traditional medicine to stimulate wound healing (Farahpour 2019; Firdous and Sautya 2018). However, studies on the potential use of certain combined extracts for surgical healing are still quite incipient. Such studies are relevant because they present an important element of innovation in both veterinary and human surgery. Three plants can be highlighted: andiroba (Carapa guianensis, Meliaceae), copaiba (Copaifera langsdorffii, Leguminosae) and guarana (Paullinia cupana, Sapindaceae), and the oils of these first two plants are used in traditional Amazonian medicine for their important pharmacological properties, including wound healing (Guimarães et al. 2016; Henriques and Penido 2014).

Andiroba's Healing Potential

Andiroba (Carapa guianensis Aubl.) is an arborial plant, belonging to the Meliaceae botanical family, found mainly in floodplains and igapós, but can also be cultivated on dry land. Its popular name derives from the term tupi ãdi'roba, which means "bitter oil." The fruit of andiroba is a globose capsule, which contains 4-6 independent internal partitions (valves). When the fruit falls, the valves separate, releasing from 4 to 12 seeds, which are brown in color and can vary greatly in shape and size (Ministry of Health 2015).

Andiroba is used in folk medicine in Brazil and other countries, covering the Amazon rainforest. Its geographical distribution extends from Central America to northern South America, including French and British Guiana, the Caribbean, Trinidad, Venezuela, Ecuador, Colombia, Peru and Brazil, where its distribution covers the northern and northeastern regions. This species is also present in western India and southern Africa (Henriques and Penido 2014; Ministry of Health 2015).

Virtually all parts of the andiroba tree are used, including seed oil, which is popularly used to treat inflammation and infections (Henriques and Penido 2014). The andiroba monograph, published by the Brazilian Ministry of Health in 2015, describes information about its chemical composition in different types of extracts and oil of this plant (Table 1) (Ministry of Health 2015).

Table 1. Bioactive molecules identified in andiroba seed extracts obtained with different solvents and in pressed oil (Reference: Brazilian Ministry of Health 2015)

Chemical classes	Seed		
	Aqueous extract	Hexanic Extract	Oil
Triterpenoids	Absent	Present	Present
Tetraterpenes	Absent	Absent	Present
Alkaloids	Absent	Absent	Present
Tannins	Absent	Absent	Absent
Limonoids	Absent	Absent	Present

Studies that chemically characterized andiroba oil describe that it is mainly composed of saponifiable material, as it has a high percentage of unsaturated fatty acids, which is of great interest to the cosmetic industry. In addition, about 2 to 5% of this oil is composed of an unsaponifiable fraction consisting of bioactive molecules called limonoids, which are tetranortriterpenoids, with emphasis on 6α-acetoxygedunine, 7-deacetoxy-7-oxogedunine, andirobine, gedunine, angolensate. methyl, epoxyiazadiradione, 6α-acetoxyhepoxyiazadiradione, 6β-acetoxygedunine, 11β-acetoxygedunine, 6α, 11β-diacetoxygedunine, 6β, 11β-diacetoxygedunine, 6α-hydroxygedunine, 17β-hydroxydiazir.

Limonoids are bioactive molecules that appear to contribute significantly to the medicinal properties of andiroba. Chemically, these molecules have moderate polarity, are insoluble in water and hexane, but are soluble in alcohol and acetone.

In fact, there are more than 300 types of limonoids described in the literature, also found in citrus fruits (Roy and Saraf 2006).

According to Henriques and Penido's review (2014), evidence from different experimental rodent models showed that andiroba oil inhibits edema formation by compromising the signaling pathways triggered by histamine, bradykinin and platelet activating factor. Tetranortriterpenoids also decreased the intensity of production of proinflammatory mediators, which trigger leukocyte infiltration at the inflammatory site, including interleukin 1-beta (IL-1β) and tumor necrosis factor-alpha (TNF-α) levels.

This phenomenon seems to depend on the inhibition of the activation of nuclear factor kappa B (NFκB). A more recent study also described that andiroba decreased macrophage activation, thus having a potential anti-inflammatory effect (Higuchi et al. 2017).

Milhomem-Paixão et al. (2016) described that andiroba oil has a high antioxidant capacity. This plant also has antimicrobial effect reported in the literature (Meccia et al. 2013). Limonoids obtained from andiroba also have a hepatoprotective effect (Brito et al. 2013; Ninomiya et al. 2016).

Considering its chemical composition based on limonoids, it is possible that the biological effects of andiroba oil are closely related to these constituents. Specifically regarding its action on the skin, a study conducted by Morikawa et al. (2018) described that andiroba oil is capable of promoting collagen synthesis in intact dermal fibroblasts from humans without causing cytotoxicity to these cells.

In addition, investigations into the effect of andiroba on healing processes have shown interesting results. This is the case of research conducted by Nayak et al. (2011) in rats using wound, excision, and dead space models. To conduct the study, the animals were divided into two groups (control and test) in each of the models tested, and were treated with andiroba leaf extract. The results showed an increased rate of wound contraction, higher resistance to skin rupture and higher hydroxyproline content in the animals treated with the extract, indicating its potential application in the healing of skin lesions.

Another study conducted on Wistar rats by Chia et al. (2018), evaluated the effects and mechanisms of topical treatment of an andiroba emulsion on skin wounds. The authors evaluated this effect 3, 7, 15 and 20 days after surgical intervention in the animals and observed an important anti-inflammatory role of andiroba in healing via increased levels of growth factor beta 3 (TGF-β3). The treated wounds also presented less dense and more organized collagen fibers. Furthermore, a study conducted by Wanzeler et al. (2018) described that andiroba oil has the property of reducing the intensity of oral mucositis, which is highly prevalent in cancer patients.

A study of the 7-deacetoxy-7-oxogedunine limonoid found in andiroba suggested that this molecule has the ability to suppress adipogenesis in adipocytes *in vitro*. This suppression occurred in the early phase of differentiation of these cells through repression of glucose uptake by them (Matsumoto et al. 2019). This work is quite interesting and it is believed that it may have some level of relevance regarding the clinical and surgical application of extracts based on andiroba oil.

As plant extracts can also have negative effects, especially genotoxicity, studies related to their safety are of great relevance. An investigation conducted by Milhomem-Paixão et al. (2016) demonstrated absence of hematotoxic, genotoxic or mutagenic effects of andiroba oil. Accordingly, the study by Araujo-Lima et al. (2018) suggested that the oil is not genotoxic, especially if it is cold extracted.

Due to the water-insoluble nature of andiroba oil, emulsions have been developed for pharmacological use. Emulsions are heterogeneous mixtures of two or more immiscible liquids, where droplets of a liquid are dispersed in a second continuous liquid phase. In fact, liquid/liquid immiscibility creates an interfacial tension between the two phases, which attributes thermodynamic instability to such systems (Pereira and Garcia-Rojas 2015).

In the production of stable emulsions, surfactant molecules are added to the system, but the most stable emulsions are obtained via emulsifiers. Ferreira et al. (2010) produced various formulations of andiroba oil-based emulsions and described them in the literature. However, the use of solvents is also realistic regarding the extraction of bioactive molecules (especially limonoids) from andiroba oil. These solvents allow the production of aqueous compounds, which could be used in the production of easily topically applied gels. Therefore, the use of solvents, such as alcohol and acetone, makes it possible to extract andiroba bioactive molecules for the production of new products with aqueous dilution capacity.

Copaíba's Healing Potential

Another Amazonian plant with healing properties is copaiba (Copaifera langsdorffii Def), which belongs to the Leguminosae family and is native to the tropical regions of Latin America and West Africa. In Brazil, this tree is found in the southeast, midwest and amazon regions (Pieri et al. 2009).

From Copaiba, an oil-resin is extracted, which can be obtained through three extraction processes. In traditional extraction, an opening is made in the tree's stem, using an ax, and the oil is collected. However, this process leads to the loss of large amounts of oil that runs down the stem, in addition to the tree being at great risk of dying and the risk of not being able to recollect from a previously extracted tree. Copaiba oil can also be obtained through total extraction, in which the trees are felled and opened. However, this process is also not productive and contributes to the deforestation of the species. Finally, there is the rational method of extraction of copaiba oil, in which a small hole is drilled in the trunk of the tree, using an auger to reach the shaft, where the oil is located. In the hole, a pipe with a hose is inserted, which conducts the oil to a container. Shortly after collection, the piece of pipe is sealed with a thread and remains in the trunk to facilitate future extractions (Romero 2007).

Copaiba oil is a transparent liquid which color varies from yellow to brown and, in the species Copaifera langsdorfii, has a red color. In chemical terms, the oil consists of sesquiterpenes and diterpenes. Studies have already identified 72 types of sesquiterpenes and 28 types of diterpenes in copaiba oils (Pieri et al. 2009; Veiga Junior and Pinto 2002).

From the biological point of view, this oil is a product of excretion or detoxification of the plant and works as a defense against animals, fungi and bacteria (Romero 2007). The oil is used as a herbal medicine, being widespread in Brazil and sold in a large number of free markets, herbalists and pharmacies of natural products (Garcia and Yamaguchi 2012). It is also used in the cosmetics, perfume (as a fixative), varnish (as a drying) and photography (as an accelerator) industry (Veiga Junior and Pinto 2002).

Copaiba oil is popularly used as a herbal medicine because it has several pharmacological properties, including wound healing, anti-inflammatory, antimicrobial, analgesic and anticancer. These properties, in particular their healing action, have been proven via scientific studies (Abrão et al. 2015; Basile et al. 1988; Garcia and Yamaguchi 2012; Gushiken et al. 2017).

Masson-Meyers et al. (2013) evaluated the cytotoxicity of copaiba oil-resin in 3T3 fibroblasts and investigated its healing effect on rabbit ear wounds. The wounds were treated topically for 21 days and evaluated on days 2, 7, 14, and 21 for healing rates and histology. The resin oil caused no cytotoxic effects on fibroblasts up to 100 µg/mL and, at appropriate doses, improved the wound healing process in rabbits.

A study conducted by Gushiken et al. (2017) investigated the effect of oil-resin and hydroalcoholic extract of Copaifera langsdorffii leaves on male Wistars rat wounds. In these animals, the wounds were treated once a day for 3, 7 or 14 days, and the wound areas were measured. The results showed macroscopic retraction of wounds treated with oil-resin or hydroalcoholic extract, and both treatments showed anti-inflammatory activity. In molecular and immunohistochemical analyzes performed on animal skin samples, both products positively induced angiogenesis, reepithelization, wound retraction and tissue remodeling. The favoring of angiogenesis by treatment with copaiba oil had been previously reported by Estevão et al. (2009) and Estevão et al. (2013).

Guarana Potential in the Quality of the Scar Process

Guarana (Paullinia cupana var. Sobilis (Mart.) Ducke) is a species native to the Amazon region, belonging to the Sapindaceae family. This plant is known to have stimulating and medicinal properties, which can be attributed to the presence of several bioactive molecules in its composition, especially caffeine, theobromine, theophylline and catechins, among others (Bittencourt et al. 2013). A study conducted by Angelo et al. (2008) evaluated the guarana transcriptome, identifying transcripts of molecules present in this plant, which are also found in coffee, tea (green and black) and chocolate, indicating the sharing of functional properties of guarana with these foods.

Among its biological properties, observed from studies in experimental models and some in humans, are: antioxidant, antimicrobial, antiallergic, antiplatelet, anti-inflammatory, hepatoprotective, gastroprotective, genoprotective, chemopreventive and anticarcinogenic effects. Guarana also provides improved cognitive performance, has antidepressant, antifadigant, antiobesogenic and hypolipidemic effect, and has aphrodisiac properties (Marques et al. 2019).

Although guarana is not traditionally used in healing processes, as it has well-established antioxidant, anti-inflammatory and antimicrobial properties, a previous study conducted by Machado et al. (2015) investigated whether this plant could have an effect on adult stem cells obtained from liposuction tissues of healthy women who underwent cosmetic surgery.

In fact, one of the most studied areas of contemporary science involves stem cells. Initially, it was thought that there were only embryonic stem cells that, if properly induced, could regenerate tissues and organs. This view changed as stem cells were discovered in adult tissues, such as adipose tissue, which could be used to regenerate tissues, including cartilage and bones. However, a major problem still to be solved by modern science is that these cells are in small quantities in the body. Thus, when we obtain them, they must first be cultivated in the laboratory and then transplanted to patients to assist in tissue regeneration. The problem is that when stem cells are cultured, they often age rapidly, losing their ability to proliferate and differentiate into specialized adult cells (Turinetto et al. 2016).

Thus, the study by Machado et al. (2015), cited above, was conducted to investigate the impact of guarana supplementation on reversal of primary stem cell senescence indicators obtained from human liposuction samples. These cells were maintained in culture conditions until the 8th pass, when they began to show senescence characteristics, such as decreased cell proliferation rate and increased indicators of oxidative stress. In 72-hour senescent stem cell cultures in which guarana (5 mg/g) was added, there was a 79% ± 15% increase in cell proliferation rate compared to the control group. Guarana decreased markers of oxidative stress, such as the total amount of reactive oxygen species (ROS), protein carbonylation, lipoperoxidation and DNA damage, as assessed by the Cometa DNA assay. Guarana also differentially modulated the levels of antioxidant enzymes and their respective genes. The results suggested that guarana supplementation could reverse some of the initial processes of cell senescence under in vitro conditions. Based on these results, the use of guarana in the formulation of herbal medicines that enhance the healing quality was postulated.

EXPERIMENTAL MODELS

In Vitro Fibroblast Experimental Model

Despite the large amount of healing potential Amazonian plants that are used in traditional medicine, the number of controlled studies that indicate their efficacy and safety is still quite low. In part, this is due to the need for studies in animal models such as rodents, and also in humans, which need approval by the respective Research Ethics Committees and are still costly. Many of these studies fail to demonstrate the effectiveness of a plant with a potential healing effect, as it is not ethically and technically feasible to conduct research involving large numbers of treatments or animals/humans.

With the prospect of broadening studies in the area of scarring, researchers developed an *in vitro* experimental model called the "*Stratch assay*" originally developed by Liang et al. (2007), with culture of fibroblasts, but that can also be applied to culture of keratinocytes.

Stratch assay is an easy, inexpensive and well-developed method for measuring fibroblast cell migration and proliferation *in vitro*. The basic steps involve creating a "scratch" pipette tip into a cell monolayer, capturing images at the beginning and at regular intervals during cell migration (Liang et al. 2007).

Compared to other methods, the *in vitro* scratch assay is particularly suited for studies on the effects of cell-matrix and cell-cell interactions on cell migration, which mimic cell migration during *in vivo* wound healing and are image compatible live cells during migration to monitor intracellular events if desired. In addition, this method has also been adopted to evaluate other healing markers, including oxidative, inflammatory and apoptotic markers, at protein level and gene expression (Cory 2011).

Californian Worms as a Model of Healing and Regeneration

As noted earlier, the development of effective therapies with wound healing potentials has been highly anticipated as significant levels of wound-induced morbidity and mortality worldwide have exerted severe economic and social damage throughout the world society (Sen et al. 2009).

To this end, many studies are conducted in animal models *in vivo*. One noteworthy model with respect to its regenerative properties concerns annelid worms (earthworms), such as the Eisenia fetida (Savigny, 1826) (Lumbricidae), popularly known as the red worm or Californian worm.

In fact, the very use of these animals in wound care has attracted the attention of scientists. As highlighted by the review by Yang et al. (2017), there is a long-recorded history in which earthworms were used for medical purposes in both Asian and North American countries. In China, doctors used discarded worms to treat a number of common ailments, including burns, arthritis, itching, and inflammation. With the development of modern science, various bioactive components with medical functions were extracted from the worm. These include glycoprotein G-90, extracted from Eisenia fetida, which has various biological activities, including antimicrobial, antitumor, antiviral, anticoagulant, antioxidant and mitogenic. The study by Yang et al. (2019) showed that G-90 protein accelerates regeneration after three days of earthworm amputation. This process is associated with G-90 protein induction of heat shock proteins such as HSP-70.

Deng et al. (2018) investigated the effect of worm extract on the wound healing process in mouse skin and the data obtained showed that the extract decreased healing time and reduced the negative effects of inflammation, thus improving the healing quality of mice wounds.

Previous studies have suggested that the worm can regenerate excised tissues and organs via activation of genes present in embryonic stem cells that remain inactive after this period in vertebrates. Some of these genes are involved in triggering metamerization (body segmentation), neurogenesis and differentiation of other tissue structures. That is, when a part of the earthworm is cut, the cells that remained in that region de-differentiate into embryonic cells, and then initiate the redifferentiation (regeneration) of the lost structures (Bhambri et al. 2018).

As the worm is an animal with great regenerative capacity, studies evaluating how certain natural or pharmacological products can influence this regeneration process are of great relevance. These analyzes can be done based only on the external morphology of the worm, investigating the regeneration rate and its quality.

Given the context presented here, the objective of this chapter is to evaluate the effect of a product, obtained from andiroba, copaíba and guarana (ACG), in the form of biphasic oil and emulsion, through the analysis of wound healing models *in vitro* (using dermal fibroblasts) and *in vivo* (using the Eisenia fetida Californian worm as a model of body regeneration via surgical removal of the caudal region).

METHODS

ACG® Product (Andiroba, Copaíba and Guaraná)

One of the problems associated with the use of andiroba (Carapa guianensis, Meliaceae) and copaiba (Copaifera langsdorffii, Leguminosae), in scarring processes, concerns the strong odor of these oils. In order to minimize this characteristic and concentrate bioactive molecules present in these oils, an extraction process and the production of a low odor oil with higher concentration of limonoids and terpenes were made from two crude oils of andiroba and copaíba, using as solvents 70% alcohol and 100% acetone and heating to reintegrate the bioactive molecules dissolved in the aqueous phase (hydrolate) to the lipophilic phase (oil). This process has led to the evaporation of the essential oils, which produce the characteristic odor of crude oils, thus reducing their detection. As these oils do not have polyphenols, which have several biological properties, such as antioxidant and anti-inflammatory action, a solution made with roasted guarana seed obtained by extraction with hot water was added to the product.

Since this process was previously developed at the Biogenomic Laboratory of the Universidade Federal de Santa Maria (UFSM), in collaboration with the Fundação Universidade Aberta da Terceira Idade (FUNATI), it is in the process of submitting for a patent. For this reason, the details of the process for obtaining ACG® biphasic oil will not be described here in order to protect the intellectual property of the inventors.

In *in vitro* studies in cell culture, to evaluate the antioxidant capacity and genotoxic potential of biphasic oil, it was previously diluted with 10% ethyl alcohol to allow its integration in oily solutions.

In *in vivo* studies with earthworms, in addition to biphasic oil, an emulsion of the product was also formulated, and both dosage forms were tested on animals. The reagents and emulsion preparation procedures were those used in the pharmaceutical industry, under the responsibility of the pharmacist Cibele Ferreira Teixeira.

ACG® Product's Antioxidant Capacity

Antioxidant capacity was assessed by testing with 2,2-diphenyl-1-picrylhydrazyl (DPPH •), which is a stable, intense purple color radical. This assay is based on the ability of substances to sequester the DPPH • radical, which can be measured by decreasing its absorbance. The percentage of unreacted DPPH • represents the percentage of antioxidant capacity (% AA) and is proportional to the concentration of antioxidant substance. The concentration that causes a decrease in initial DPPH • concentration by 50% is defined as either effective concentration (EC50) or inhibitory concentration (IC50). The test was performed by spectrophotometry, according to Zhang et al. (2007).

Genotoxicity/Genoprotection of ACG® Product

The Gemo test is an *in vitro* method that assesses the genomodifying ability of chemical and synthetic compounds. This protocol does not use biological systems and was developed and used for the quantification of double-stranded DNA (dsDNA) exposed to chemicals. The method is performed in a 96-well black plate using a highly specific dsDNA (PicoGreen®) dye and purified calf thymus dsDNA. The test includes a reference pro-oxidant, hydrogen peroxide (3 molar H_2O_2), which allows the comparative analysis of the obtained data. It is possible to classify the tested substance in several levels of genotoxicity and also to indicate if it has potential for genoprotection (Cadoná 2013).

In Vitro Protocol with Fibroblasts

The experimental part of the study was conducted at the premises of the UFSM Biogenomic Laboratory, which has already developed several studies in partnership with Amazonian research institutions.

HFF-1 (ATCC® SCRC-1041 ™) strain fibroblasts, thawed by the Rio de Janeiro Cell Bank, were grown in Dulbecco's Modified Eagle Medium (DMEM) medium supplemented with 15% fetal bovine serum, 1% antibiotics (penicillin/streptomycin) and 1% antifungal (amphotericin B). The cells remained in cell culture under sterile conditions in an appropriate incubator at 37°C with 5% CO_2 saturation until the amount of cells needed to perform the experiments was obtained.

For *in vitro* testing, ACG was dissolved in 1% Dimethylsulfoxide (DMSO) and a Scratch assay was performed by Nicolaus et al. (2017) with minor modifications. Briefly, the cells were seeded at a density of 1×10^4 cells/well in a 24-well culture plate and incubated overnight. After the incubation, DMEM culture medium was aspirated and the adherent cell layer was scratched with a sterile yellow pipette tip. Further, cell debris were removed by rinsing with PBS. The complete medium with and without the hydroalcoholic extract of barbatimão was then added to each well. The image of the scratched area was captured under bright field microscopy (20×) and the migration was analyzed using Digimizer software (version 5.3.4) which allowed manual measurements as well as automatic object detection with measurements of cell culture characteristics. The wound area was measured using controller software.

Teared fibroblasts were treated with the product at a concentration of 2 µL/mL and then the effect of ACG on the migration of these cells and on the modulation of proliferative, oxidative and inflammatory markers was evaluated.

Trials Performed in the *In Vitro* Study

Cell proliferation indicators: Protein levels of two growth factors were quantified after 72 hours of treatment: fibroblast growth factor (FGF-1), involved in stimulating DNA synthesis and proliferating a wide variety of cells, and keratinocyte growth factor (KGF), also known as FGF-7, which induces mitogenic and cell survival activities. Quantification of these two markers was performed using the Enzyme-Linked Immunosorbent Assay (ELISA) enzyme-linked immunosorbent assay using the Abcam® kit as recommended by the manufacturer. The sensitivity and detection range of FGF-1 and KGF were, respectively, 0.78-50 ng/mL and 25-1600 ng/mL.

Indicators of oxidative stress: Quantitation of ROS after 24 and 72 hours of cell treatment with ACG was performed by the 2',7'-Dichlorofluorescein Diacetate (DCFH-DA) assay, which mainly measures the presence of H_2O_2 (Halliwell and Whiteman 2004). Protein levels of the enzymes superoxide dismutase, catalase and glutathione peroxidase were measured by ELISA after 24 and 72 hours of treatment using the Abcam® kit, according to the manufacturer's recommendations.

Inflammatory markers: Protein levels of pro-inflammatory cytokines, IL-1β, IL-6, TNF-α and interferon-gamma (IFN-γ), as well as anti-inflammatory cytokine IL-10, were measured by ELISA after 72 hours of treatment using the Quantikine® kit and the experimental procedures followed the manufacturer's recommendations.

All immunoassays were courtesy of the Labimed Clinical Analysis Laboratory, where the analyzes were performed.

In Vivo Protocol: Eisenia Fetida Model

Californian worms were commercially acquired for the experiments. They were acclimated for 10 days under laboratory conditions, where they were kept on commercially acquired land, containing 50% soil and 50% humus. To this substrate was added 5% dry manure as a source of nutrients for the earthworms. The substrate was kept in a closed container with 60% humidity, pH 7.2-7.6, average temperature of $25 \pm 2°C$ and 12/12 hours light/dark cycle.

The worms were divided into the following treatment groups: hydrophilic negative control (water), lipophilic negative control (mineral oil), ACG biphasic oil, emulsion without the addition of ACG product and added emulsion of ACG. The concentration of the product used to perform this protocol was the same as that chosen in the *in vitro* study.

The experimental design was conducted with the distribution of 5 worms for each treatment, with repetition of three times. The study was performed from the surgical removal of the caudal end of the adult wormed worm (~ 1 cm), followed by treatment and regenerative pattern analysis after 7 days. The protocol used was similar to that of Tao et al. (2018) and is described below:

1. The worms were removed from the earth and washed with plenty of filtered water. They were then lightly dried using paper towels;
2. Soon after, the worms were dipped (~ 10 seconds) in a 10% ethyl alcohol solution, placed in a petri dish, to aid in the asepsis of the animals and also to produce an exothermic reaction, which enhances body cooling. and temporary paralysis;
3. The worms were then transferred to a frozen water Petri dish to complete the cryanesthesia process;
4. Steps (2) and (3) ensure that the worm does not curl and become immobilized, thus allowing surgical caudal removal (last 5 segments) with the aid of a scalpel;

5. Immediately after surgical caudal removal, 2 µl or 2 µg (in the case of emulsion) of the respective treatment were placed on the wound with the aid of a pipette or brush (in the case of the emulsion);
6. The earthworms were then transferred to earthen containers and kept under the same conditions as described above;
7. The removal area of each worm was photographed using a magnifying glass with integrated Leica® brand photo system;
8. After 7 days, the effect of the ACG product on the occurrence and quality of tissue regeneration in earthworms was analyzed.

Quality Markers of Tissue Regeneration in Eisenia Fetida Visually Analyzed

- Occurrence and evolution of regeneration;
- Myelinization of regenerated segments: In the whole earthworm, a clear separation between the segments can be observed, made by dark (wider) and light (thinner) bands;
- Presence of ventral bristles in regenerated segments: The earthworm has two bristles in each segment, which aid in locomotion and excavation of the animal. These bristles are located only in the ventral region of the worm;
- Presence of vascularization (angiogenesis of the great vessels) in the regenerated area: In the earthworm, there is a large dorsal vessel and a large ventral vessel, which branch into capillaries. These vessels will need to be regenerated after surgical removal;
- Maintenance of proportionality, in terms of width, of the regenerated segments, in relation to the other body segments. The most tapering regeneration of the last caudal segment was also considered in the analysis.

Statistical Analysis

All *in vitro* and *in vivo* tests were performed in triplicate. The values obtained were presented as a percentage in relation to the negative control value (without any treatment). Results were presented as mean ± standard deviation (SD) or standard error (SE). The treatments were compared by analysis of variance (one or two ways, as appropriate), followed by Dunnet, Tukey or Bonferroni post hoc tests (as appropriate) and the software used was Graph Pad Prism, version 5.0. Analyzes of the earthworm regenerated areas were made from visual observation and evaluation via Digimizer image analysis software package (version 5.3.5, last modified January 2019, MedCalc Software, Belgium), which allows measurements as well as automatic object detection with object characteristic measurements. Considering that the average size of the regeneration perimeter is variable, the comparison between the treatment groups was made through the medians, using the nonparametric Kruskal-Wallis analysis. All p values were two-tailed. The alpha value was set to <0.05 to determine statistical relevance.

RESULTS

The antioxidant capacity of the ACG product was evaluated in comparison to rutin polyphenol, used as a reference molecule. The results showed that the ACG product presented higher antioxidant capacity than pure oils and guarana aqueous extract, when analyzed separately. The antioxidant capacity of the product was closer to that observed in rutin, although significantly lower (Figure 1).

Genoprotective capacity was also evaluated and the results were similar to those observed in antioxidant capacity. The analysis of these results made it possible to choose the concentration range of the ACG product to be tested in fibroblasts, which was 0.1; 0.3; 0.5; 1.0 and 2.0 μL/mL. The concentration of 1 μL/mL did not cause mortality in 24-hour cultures and, on the contrary, significantly increased viability compared to the control group (114.5 ± 2.3% (p < 0.001)).

Based on these initial results, the concentration of 2 µL/mL of the ACG product was chosen for the testing of its potential healing effect.

Stratch assay was then performed on fibroblasts and, as can be seen in Figure 2, ACG treatment was able to induce a higher rate of fibroblast migration when compared to torn and untreated cells.

In torn fibroblast cultures treated for 72 hours with ACG, the levels of two cell proliferation markers were analyzed: FGF-1 and KGF (Figure 3). ACG treatment elevated the protein levels of these two growth factors.

Oxidative markers analysis included quantification of ROS levels by the DCFH-DA technique after 24 and 72 hours of treatment of the torn cells. However, in this analysis, no significant difference was observed between cells treated and untreated with ACG (Figure 4).

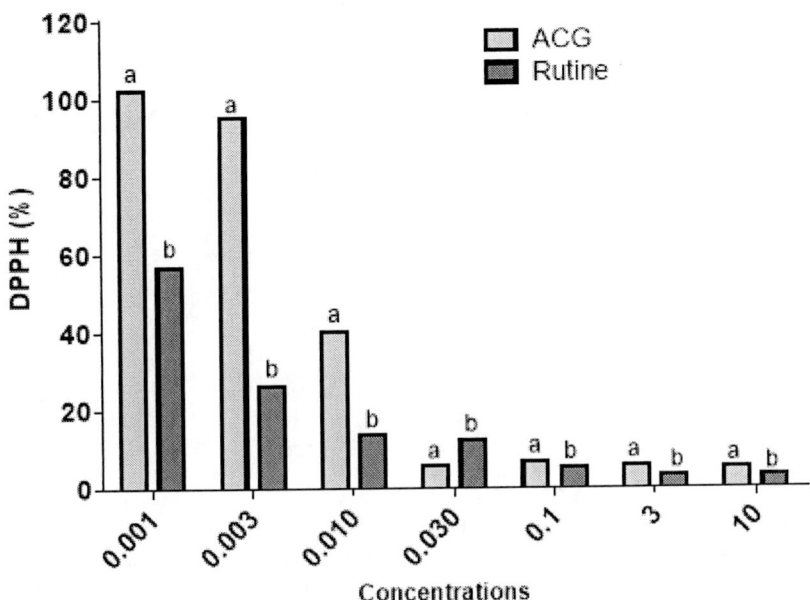

Figure 1. ACG antioxidant capacity, analyzed via DPPH test •.

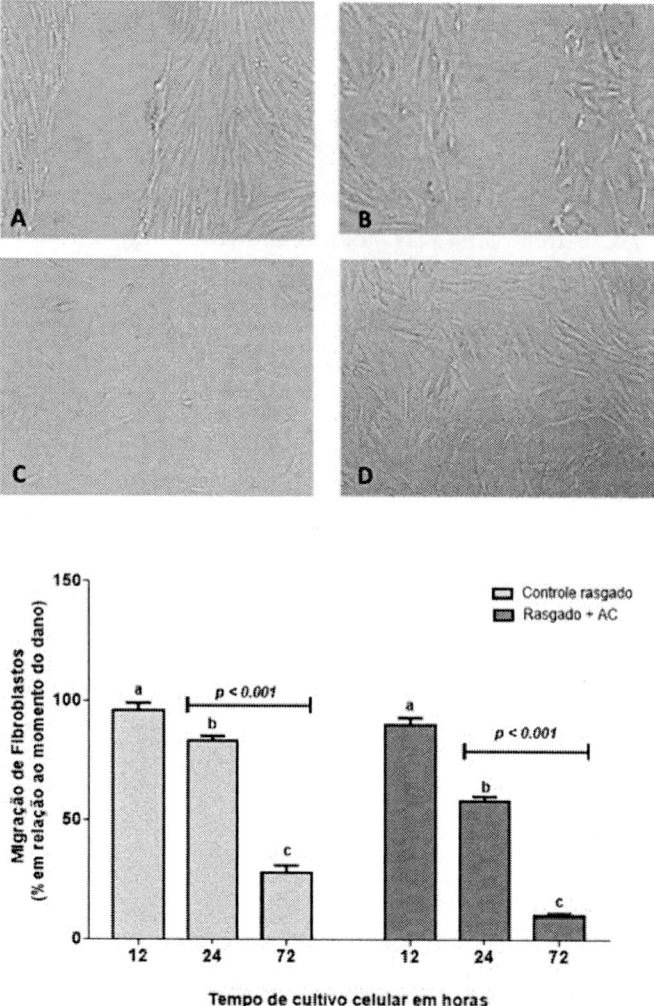

Figure 2. Analysis of fibroblast migration using the assay *Stratch in vitro*: (A) Strach of the fibroblastos culture; (B) Fibroblast migration after 12 hours of strach in untreated culture; (C) Fibroblast migration after 24 hours of strach in untreated culture; (D) Migration of fibroblasts after 24 hours of streaching in culture treated with ACG; (E) Fibroblast migration from the tear wound healing assay, expressed as relative cell migration, calculated by dividing the percentage change in the strach area of ACG-treated cells by 12h, 24h, or 72h from the time of tear (time 0). Results between control and ACG were compared by analysis of variance One-Way, followed by *post hoc* test *Tukey*. Statistically significant differences are denoted by different letters. (a, b, c).

Analysis of the Effect of an Andiroba, Copaíba and Guaraná ...   91

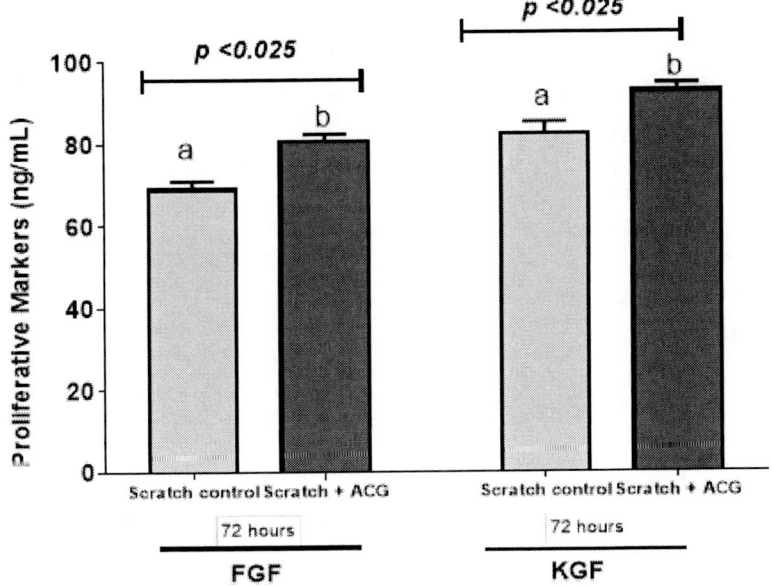

Figure 3. FGF-1 (Fibroblast Growth Factor) and KGF (Keratinocyte Growth Factor) proliferative markers, comparison between untreated and ACG-torn control groups. Data are presented as mean values ± standard error and expressed as a percentage of the torn control value. Different letters indicate statistical differences between control and ACG by one-way ANOVA analysis, followed by Tukey's post hoc test with $p < 0,05$.

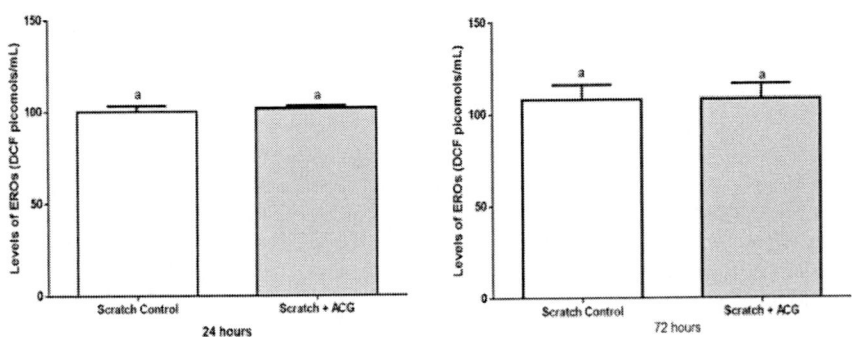

Figure 4. Levels of Reactive Oxygen Species, after 24 and 72 hours of ACG treatment. Data are presented as mean values ± standard error. Different letters indicate statistical differences between control and ACG by one-way ANOVA analysis, followed by Tukey's post hoc test with $p < 0,05$.

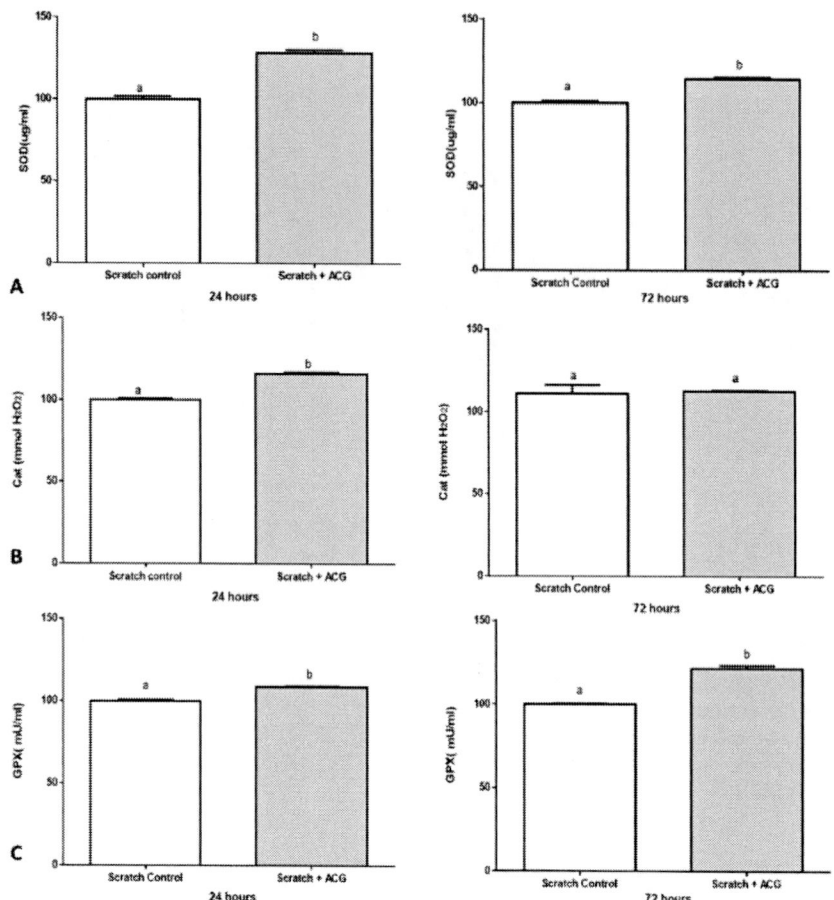

Figure 5. Protein Levels of Antioxidant Enzymes Superoxide Dismutase (SOD), Catalase (CAT) and Glutathione Peroxidase (GPX): (A) Protein levels of SOD; (B) Protein levels of CAT; (C) Protein levels of GPX. Results are presented comparing the levels of each enzyme in untreated and treated scratch cells after 24 and 72 hours. Data are presented as mean values ± standard error. Different letters indicate statistical differences between control and ACG by one-way ANOVA analysis followed by *post hoc* de *Tukey* with p < 0,05.

The protein levels of the antioxidant enzymes superoxide dismutase, catalase and glutathione peroxidase were also analyzed. The results, after 24 and 72 hours of exposure of torn fibroblasts to the ACG, are presented in Figure 5.

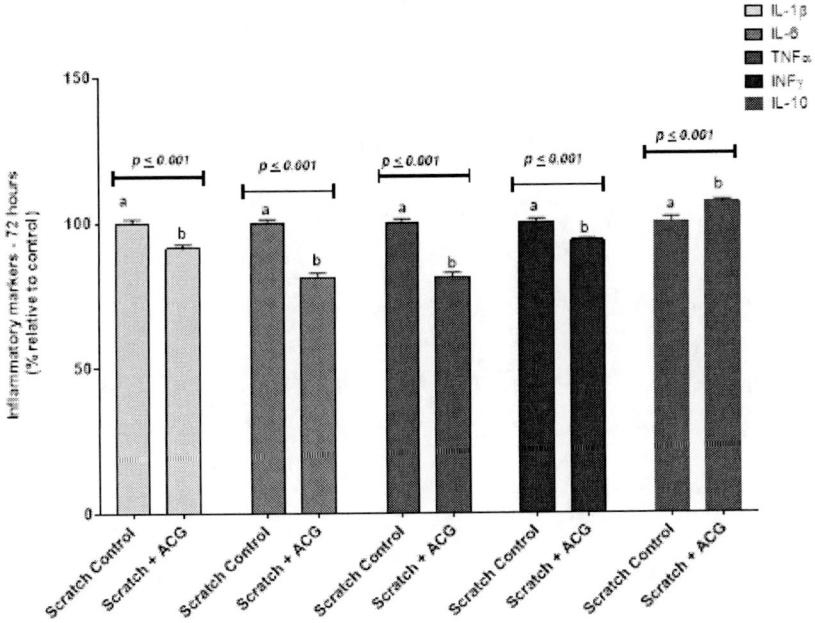

Figure 6. Markers of the inflammatory metabolism of torn fibroblasts. Compound Exposure ACG, for 72 hours, decreased all proinflammatory cytokines (IL-1β, IL-6, TNF-α, IFN-γ) and increased anti-inflammatory cytokine (IL-10). Data are presented as mean values ± standard error. Different letters indicate statistical differences between control and ACG by one-way ANOVA analysis, followed by *Tukey's post hoc tes*t with p<0,05.

After 72 hours of treatment with the ACG compound, protein levels of the pro-inflammatory cytokines IL-1β, IL-6, TNF-α, IFN-γ, and anti-inflammatory cytokine IL-10 were also evaluated. Again, the compound was pro-healing by modulating anti-inflammatory activity by decreasing proinflammatory cytokine levels and raising anti-inflammatory cytokine IL-10 levels compared to untreated cells (Figure 6).

The effect of ACG product on the regeneration of the caudal region of worms submitted to surgical removal was evaluated. Figure 7 shows an illustration of the surgical removal process performed on animals.

The ACG product was tested for two pharmaceutical presentation forms, biphasic oil (oily/aqueous) and emulsion.

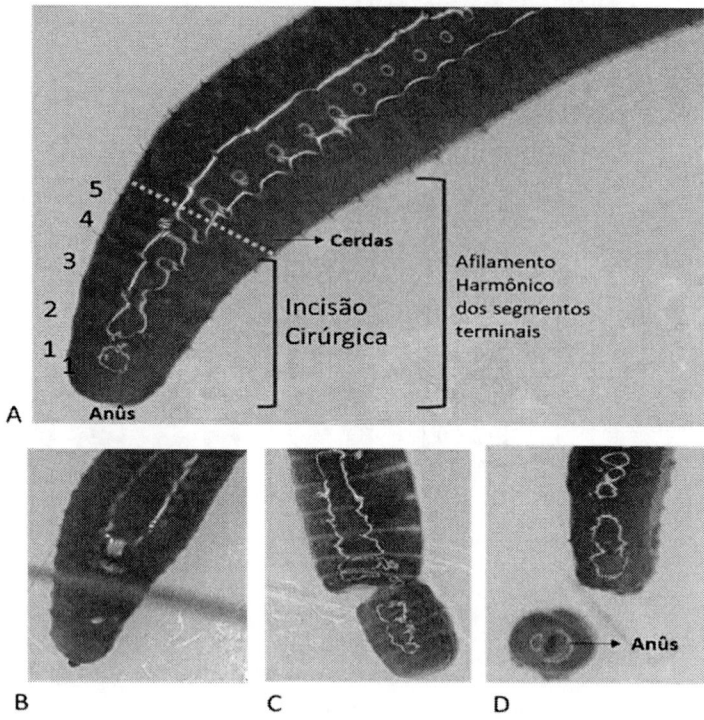

Figure 7. Main stages of surgical removal of the caudal region in earthworms: (A) Earthworm with intact posterior part, highlighting the 5 segments that were surgically removed. The photo also shows the occurrence of harmonic tapering of the last segments in relation to the anterior part of the worm. The bristles, present in each segment, are also identified; (B) Surgical scalpel removal of the 5 segments of the caudal region; (C) Detail of the separation of segments from the rest of the body; (D) Removal Completely.

As can be seen in Figure 8A, seven days after the surgical removal of the caudal region, the control worms presented a partially complete regeneration, with nascent bristles, segments in the early phase of myelination and separation in light and dark bands. The larger blood vessels were still thin and of low visibility. The proportion of regenerated segments relative to the others was maintained even though the final segment was no longer tapered than the others (Figure 8B).

In the presence of ACG biphasic oil, regeneration was complete, with well-defined dark and light segments and the presence of bristles in the ventral region (Figure 8C). The vascularization of the regenerated region was also visible and within the expected morphological pattern (Figure 8D). The regenerated segments presented proportionality in relation to the others, and the last segment already had a thinner aspect (Figure 8E).

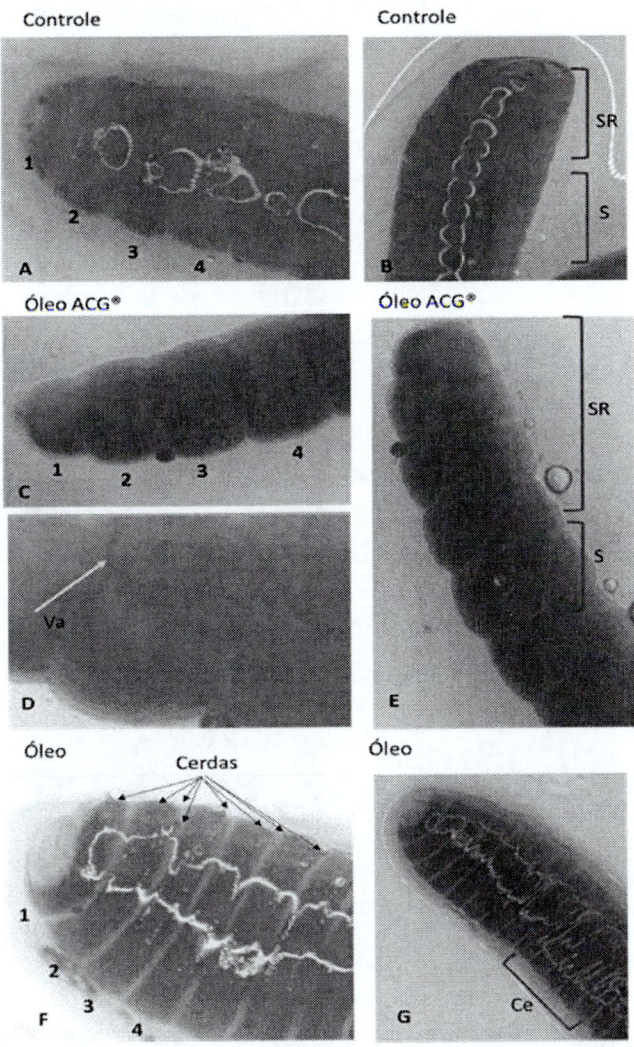

Figure 8. Comparison of the earthworm regenerated areas of the hydrophilic negative control (water), ACG biphasic oil (produced from andiroba, copaiba and aqueous extract of guarana powder) and mineral oil (lipophilic negative control) groups. SR = regenerated segments; S = other segments; Va = blood vessels.

When earthworms were treated with mineral oil, the segments presented hypermyelination and the presence of extra pairs of bristles in the dorsal region of the 1st segment and between the 2nd and 3rd segments. Two additional pairs of bristles were also identified in the dorsal region of the last segment anterior to the cut site, suggesting that the mineral oil also induced morphological changes in this segment that was not removed but was injured as a result of surgical removal. Ventral bristles continued to be observed in earthworms treated with oil in segments that were not surgically removed. Treatment with mineral oil did not induce changes in the proportions of the regenerated segments in relation to the others, and the last segment tapering in these worms was observed, indicating complete regeneration of the surgically removed area (Figures 8F, 8G).

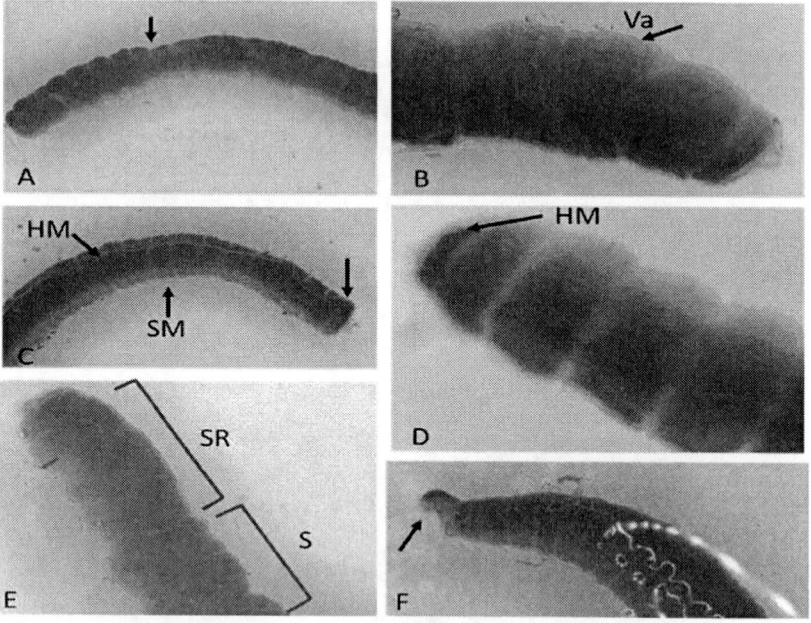

Figure 9. Representation of the main changes observed in earthworms submitted to surgical removal of the caudal region and treated with emulsion (with and without the ACG product). SR = regenerated segments; S = other segments; Va = blood vessels; MH = hypermyelination; SM = no myelination.

Contrary to the expected beneficial effect on the quality of the earthworm flow regeneration, the emulsion, with or without the presence of the ACG product, caused extensive changes in this process, as shown in photographs in Figure 9. These include: (1) changes in the proportions (widths) between the regenerated segments and the others, often resulting in clear morphological disorganization, with changes in the convex regions, which should be observed mainly in the dorsal region (Figures 9A-9E); (2) changes in the last caudal segment, which often did not present the usual tapering (Figures 9B, 9C), were hypermyelinated in relation to the other regenerated segments (Figure 9D) and bilobulated (Figure 9F); (3) occurrence of changes in the angiogenesis pattern, with blood vessels arranged outside the usual pattern (Figure 9B); (4) occurrence of heterogeneous myelination, with hypermyelination in the dorsal region (MH) and without myelination in the ventral region (MS) (Figure 9C). Ce = Bristles.

CONCLUSION

Despite the methodological limitations inherent to *in vitro* work and non-mammalian experimental models, we consider the data obtained to be relevant, since these initial results show antioxidant action, anti-inflammatory and pro-healing of the developed compound. More studies need to be carried out to prove the results obtained here, however, we believe that in the future the development of a commercial product for clinical purposes may be a reality, benefiting patients, health professionals and also the productive chains of the Amazon.

REFERENCES

Abrão, Fariza., Costa, Luciana D. A., Alves, Jacqueline M., Senedese, Juliana M., Castro, Pâmela T., Ambrósio, Sérgio R., Veneziani, Rodrigo

C. S., Bastos, Jairo K., Tavares, Denise C., and Martins, Carlos H. G. 2015. "*Copaifera langsdorffii* Oleoresin and its Isolated Compounds: Antibacterial Effect and Antiproliferative Activity in Cancer Cell Lines." *BMC Complementary and Alternative Medicine* 15:443. doi: 10.1186/s12906-015-0961-4.

Amirkia, Vafa, and Heinrich, Michael. 2015. "Natural Products and Drug Discovery: A Survey of Stakeholders in Industry and Academia." *Frontiers in Pharmacology* 6:237. doi: 10.3389/fphar.2015.00237.

Angelo, Paula C. S., Nunes-Silva, Carlos G., Brígido, Marcelo M., Azevedo, Juliana S. N., Assunção, Enedina N., Sousa, Alexandra R. B., Patrício, Fernando J. B., Rego, Mailson M., Peixoto, Jean C. C., Oliveira Jr, Waldesse P., Freitas, Danival V., Almeida, Elionor R. P., Viana, Andréya M. H. A., Souza, Ana F. P. N., Andrade, Edmar V. et al. 2008. "Guarana (*Paullinia cupana var. sorbilis*), an Anciently Consumed Stimulant from the Amazon Rain Forest: The Seeded-Fruit Transcriptome." *Plant Cell Reports* 27:117-124. doi: 10.1007/s00299-007-0456-y.

Araujo-Lima, Carlos F., Fernandes, Andreia S., Gomes, Erika M., Oliveira, Larisse L., Macedo, Andrea F., Antoniassi, Rosemar., Wilhelm, Allan E., Aiub, Claudia A. F., and Felzenszwalb, Israel. 2018. "Antioxidant Activity and Genotoxic Assessment of Crabwood (Andiroba, *Carapa guianensis* Aublet) Seed Oils." *Oxidative Medicine and Cellular Longevity* 2018:3246719. doi:10.1155/2018/3246719.

Balbino, Carlos A., Pereira, Leonardo M., and Curi, Rui. 2005. "Mecanismos Envolvidos na Cicatrização: Uma Revisão." *Brazilian Journal of Pharmaceutical Sciences* 41:27-51.

Baroudi, Ricardo. 2010. "A Segurança nas Cirurgias Estéticas Combinadas." *Revista Brasileira de Cirurgia Plástica* 25:581-582. doi: 10.1590/S1983-51752010000400002.

Barr, Simon P., Topps, Ashley R., Barnes, Nicola L. P., Henderson, Julia R., Hignett, Susan., Teasdale, Rebecca L., McKenna, A., Harvey, James R., and Kirwan, Cliona C., On behalf of the Northwest Breast Surgical Research Collaborative. 2016. "Infection Prevention in Breast Implant Surgery - A Review of the Surgical Evidence, Guidelines and a

Checklist." *European Journal of Surgical Oncology (EJSO)* 42:591-603. doi: 10.1016/j.ejso.2016.02.240.

Barreiro, Eliezer J., and Bolzani, Vanderlan S. 2009. "Biodiversity: Potential Source for Drug Discovery." *Química Nova* 32:679-688. doi: 10.1590/S0100-40422009000300012.

Basile, Aulus C., Sertié, Jayme A. A., Freitas, P .C. D., and Zanini, Antônio C. 1988. "Anti-inflammatory Activity of Oleoresin from Brazilian *Copaifera*." *Journal of Ethnopharmacology* 22:101-109. doi: 10.1016/0378-8741(88)90235-8.

Berman, Brian, Maderal, Andrea, and Raphael, Brian. 2017. "Keloids and Hypertrophic Scars: Pathophysiology, Classification, and Treatment." *Dermatologic Surgery* 43:S3-S18. doi: 10.1097/DSS.0000000000000819.

Bhambri, Aksheev., Dhaunta, Neeraj., Patel, Surendra S., Hardikar, Mitali., Bhatt, Abhishek., Srikakulam, Nagesh., Shridhar, Shruti., Vellarikkal, Shamsudheen., Pandey, Rajesh., Jayarajan, Rijith., Verma, Ankit., Kumar, Vikram., Gautam, Pradeep., Khanna, Yukti., Khan, Jameel A., Fromm, Bastian., Peterson, Kevin J., Scaria, Vinod., Sivasubbu, Sridhar., and Pillai, Beena. 2018. "Large Scale Changes in the Transcriptome of *Eisenia fetida* During Regeneration." *PLoS One* 13:e0204234. doi: 10.1371/journal.pone.0204234.

Bittencourt, Leonardo S., Machado, Denise C., Machado, Michel M., dos Santos, Greice F. F., Algarve, Thaís D., Marinowic, Daniel R., Ribeiro, Euler E., Soares, Félix A. A., Barbisan, Fernanda., Athayde, Margareth L., and da Cruz, Ivana B. M. 2013. "The Protective Effects of Guaraná Extract (*Paullinia cupana*) on Fibroblast NIH-3T3 Cells Exposed to Sodium Nitroprusside." *Food and Chemical Toxicology* 53:119-125. doi: 10.1016/j.fct.2012.11.041.

Brasil. Ministério da Saúde. Secretaria de Atenção à Saúde. Departamento de Atenção Básica. 2012. "Práticas Integrativas e Complementares: Plantas Medicinais e Fitoterapia na Atenção Básica." Brasília: Ministério da Saúde.

Brasil. Ministério da Saúde. Secretaria de Ciência, Tecnologia e Insumos Estratégicos. Departamento de Assistência Farmacêutica. 2016.

"Política e Programa Nacional de Plantas Medicinais e Fitoterápicos." Brasília: Ministério da Saúde.

Brito, Nathalya B., Souza Junior, Jorge M., Leão, Layra R. S., Brito, Marcus V. H., Rêgo, Amália C. M., and Medeiros, Aldo C. 2013. "Effects of Andiroba (*Carapa guianensis*) Oil on Hepatic Function of Rats Subjected to Liver Normothermic Ischemia and Reperfusion." *Revista do Colégio Brasileiro de Cirurgiões* 40:476-479. doi: 10.1590/S0100-69912013000600010.

Broughton, George., Janis, Jeffrey E., and Attinger, Christopher E. 2006. "The Basic Science of Wound Healing". *Plastic and Reconstructive Surgery* 117(7S):12S-34S. doi: 10.1097/01.prs.0000225430.42531.c2.

Cadoná, Francine C. 2013. "Desenvolvimento de um Método de Análise *In Vitro* da Capacidade Genomodificadora de Compostos Químicos e Sintéticos." MSc diss., Federal University of Santa Maria.

Calixto, João B., and Siqueira Jr, Jarbas M. 2008. "The Drug Development in Brazil: Challenges." *Gazeta Médica da Bahia* 78:98-106.

Chia, Chang Y., Medeiros, Andréia D., Corraes, André M. S., Manso, José E. F., Silva, César S. C., Takiya, Christina M., and Vanz, Ricardo L. 2018. "Healing Effect of Andiroba-Based Emulsion in Cutaneous Wound Healing via Modulation of Inflammation and Transforming Growth Factor Beta 3[1]." *Acta Cirurgica Brasileira* 33:1000-1015. doi: 10.1590/s0102-865020180110000007.

Cory, Giles. 2011. "Scratch-Wound Assay." *Methods in Molecular Biology* 769:25-30. doi: 10.1007/978-1-61779-207-6_2.

Deng, Zhen-han., Yin, Jian-jian., Luo, Wei., Kotian, Ronak N., Gao, Shan-shan., Yi, Zi-qing., Xiao, Wen-feng., Li, Wen-ping., and Li, Yu-sheng. 2018. "The Effect of Earthworm Extract on Promoting Skin Wound Healing." *Bioscience Reports* 38: BSR20171366. doi: 10.1042/BSR20171366.

Estevão, Lígia R. M., Medeiros, Juliana P., Scognamillo-Szabó, Márcia V. R., Baratella-Evêncio, Liriane., Guimarães, Ednaldo C., Câmara, Cláudio A. G., and Evêncio-Neto, Joaquim. 2009. "Neoangiogenesis of Skin Flaps in Rats Treated with Copaiba Oil." *Pesquisa Agropecuária Brasileira* 44:406-412. doi: 10.1590/S0100-204X2009000400011.

Estevão, Lígia R. M., Medeiros, Juliana P., Baratella-Evêncio, Liriane., Simões, Ricardo S., Mendonça, Fábio S., and Evêncio-Neto, Joaquim. 2013. "Effects of the Topical Administration of Copaiba Oil Ointment (*Copaifera langsdorffii*) in Skin Flaps Viability of Rats." *Acta Cirúrgica Brasileira* 28:863-869. doi: 10.1590/S0102-86502013001200009.

Farahpour, Mohammad R. 2019. "Medicinal Plants in Wound Healing." In *Wound Healing - Current Perspectives*, edited by Kamil H. Dogan, 33-47. London: IntechOpen. doi: 10.5772/intechopen.80215.

Ferreira, Magda R. A., Santiago, Rosilene R., de Souza, Tatiane P., Egito, Eryvaldo S. T., Oliveira, Elquio E., and Soares, Luiz A. L. 2010. "Development and Evaluation of Emulsions from *Carapa guianensis* (Andiroba) Oil." *AAPS PharmSciTech* 11.1383-1390. doi: 10.1208/s12249-010-9491-z.

Firdous, Sayeed M., and Sautya, Dippayan. 2018. "Medicinal Plants with Wound Healing Potential." *Bangladesh Journal of Pharmacology* 13:41-52. doi: 10.3329/bjp.v13i1.32646.

Garcia, Rosangela F., and Yamaguchi, Miriam H. 2012. "Copaiba Oil and its Medicinal Properties: A Bibliographical Review." *Revista Saúde e Pesquisa* 5:137-146.

Guilhermino, Jislaine F., Siani, Antonio C., Quental, Cristiane., and Bomtempo, José Vitor. 2012. "Desafios e Complexidade para Inovação a partir da Biodiversidade Brasileira." *Revista de Pesquisa e Inovação Farmacêutica* 4:18-30.

Guimarães, A. L., Cunha, E. A., Matias, F. O., Garcia, P. G., Danopoulos, P., Swikidisa, R., Pinheiro, V. A., and Nogueira, R. J. 2016. "Antimicrobial Activity of Copaiba (*Copaifera officinalis*) and Pracaxi (*Pentaclethra macroloba*) Oils Against *Staphylococcus aureus*: Importance in Compounding for Wound Care." *International Journal of Pharmaceutical Compounding* 20:58-62.

Gushiken, Lucas F. S., Hussni, Carlos A., Bastos, Jairo K., Rozza, Ariane L., Beserra, Fernando P., Vieira, Ana J., Padovani, Carlos R., Lemos, Marivane., Polizello Junior, Maurilio., Silva, Jonas J. M., Nóbrega, Rafael H., Martinez, Emanuel R. M., and Pellizzon, Cláudia H. 2017. "Skin Wound Healing Potential and Mechanisms of the Hydroalcoholic

Extract of Leaves and Oleoresin of *Copaifera langsdorffii* Desf. Kuntze in Rats." *Evidence-Based Complementary and Alternative Medicine* 2017:6589270. doi: 10.1155/2017/6589270.

Halliwell, Barry, and Whiteman, Matthew. 2004. "Measuring Reactive Species and Oxidative Damage *In Vivo* and In Cell Culture: How Should You do It and What do the Results Mean?" *British Journal of Pharmacology* 142:231-255. doi: 10.1038/sj.bjp.0705776.

Hammond, Dennis C., and Kim, Kuylhee. 2016. "The Short Scar Periareolar Inferior Pedicle Reduction Mammaplasty: Management of Complications." *Clinics in Plastic Surgery* 43:365-372. doi: 10.1016/j.cps.2015.12.010.

Henriques, Maria G., and Penido, Carmen. 2014. "The Therapeutic Properties of *Carapa guianensis*." *Current Pharmaceutical Design* 20:850-856. doi: 10.2174/13816128113199990048.

Higuchi, Keiichiro, Miyake, Teppei, Ohmori, Shoko, Tani, Yoshimi, Minoura, Katsuhiko, Kikuchi, Takashi, Yamada, Takeshi and Tanaka, Reiko. 2017. "Carapanosins A–C from Seeds of Andiroba (*Carapa guianensis*, Meliaceae) and Their Effects on LPS-Activated NO Production." *Molecules* 22:502. doi: 10.3390/molecules22030502.

Ho, Jasmine, Walsh, Claire, Yue, Dominic, Dardik, Alan, and Cheema, Umber. 2017. "Current Advancements and Strategies in Tissue Engineering for Wound Healing: A Comprehensive Review." *Advances in Wound Care (New Rochelle)* 6:191-209. doi: 10.1089/wound.2016.0723.

Liang, Chun C., Park, Ann Y., and Guan, Jun L. 2007. "*In Vitro* Scratch Assay: A Convenient and Inexpensive Method for Analysis of Cell Migration *In Vitro*" *Nature Protocols* 2:329-333. doi: 10.1038/nprot.2007.30.

Liechty, Kenneth W., Kim, Heung B., Adzick, N. S., and Crombleholme, Timothy M. 2000. "Fetal Wound Repair Results in Scar Formation in Interleukin-10-Deficient Mice in a Syngeneic Murine Model of Scarless Fetal Wound Repair." *Journal of Pediatric Surgery* 35:866-873. doi: 10.1053/jpsu.2000.6868.

Mabona, Unathi, and Van Vuuren, Sandy F. 2013. "Southern African Medicinal Plants Used to Treat Skin Diseases." *South African Journal of Botany* 87:175-193. doi: 10.1016/j.sajb.2013.04.002.

Machado, Alencar K., Cadoná, Francine C., Azzolin, Verônica F., Dornelles, Eduardo B., Barbisan, Fernanda., Ribeiro, Euler E., Cattani, Maria F. M., Duarte, Marta M. M. F., Saldanha, José R. P., and da Cruz, Ivana B. M. 2015. "Guaraná (*Paullinia cupana*) Improves the Proliferation and Oxidative Metabolism of Senescent Adipocyte Stem Cells Derived from Human Lipoaspirates." *Food Research International* 67:426-433. doi: 10.1016/j.foodres.2014.11.056.

Mahedia, Monali, Shah, Nilay, and Amirlak, Bardia. 2016. "Clinical Evaluation of Hyaluronic Acid Sponge with Zinc versus Placebo for Scar Reduction after Breast Surgery." *Plastic and Reconstructive Surgery – Global Open* 4:e791. doi: 10.1097/GOX.0000000000000747.

Marques, Leila L. M., Ferreira, Emilene D. F., Paula, Mariana N., Klein, Traudi., and Mello, João C. P. 2019. "*Paullinia cupana*: A Multipurpose Plant – A Review." *Brazilian Journal of Pharmacognosy* 29:77-110. doi: 10.1016/j.bjp.2018.08.007.

Masson-Meyers, Daniela S., Andrade, Thiago A. M., Leite, Saulo N., and Frade, Marco A. C. 2013. "Cytotoxicity and Wound Healing Properties of *Copaifera langsdorffii* Oleoresin in Rabbits." *International Journal of Natural Product Science* 3:10-20.

Matsumoto, Chihiro, Koike, Atsushi, Tanaka, Reiko, and Fujimori, Ko. 2019. "A Limonoid, 7-Deacetoxy-7-Oxogedunin (CG-1) from Andiroba (*Carapa guianensis*, Meliaceae) Lowers the Accumulation of Intracellular Lipids in Adipocytes via Suppression of IRS-1/Akt-Mediated Glucose Uptake and a Decrease in GLUT4 Expression." *Molecules* 24:1668. doi: 10.3390/molecules24091668.

Meccia, Gina, Quintero-Rincón, Patricia, Rojas, Luis B, Usubillaga, Alfredo, Velasco, Judith, Diaz, Tulia, Diaz, Clara, Velásquez, Jesús, and Toro, Maria E. 2013. "Chemical Composition of the Essential Oil from the Leaves of *Carapa guianensis* Collected from Venezuelan Guayana and the Antimicrobial Activity of the Oil and Crude Extracts." *Natural Product Communications* 8:1641-1642.

Milhomem-Paixão, Susana S. R., Fascineli, Maria L., Roll, Mariana M., Longo, João P. F., Azevedo, Ricardo B., Pieczarka, Julio C., Salgado, Hugo L. C., Santos, Alberdan S., and Grisolia, Cesar K. 2016. "The Lipidome, Genotoxicity, Hematotoxicity and Antioxidant Properties of Andiroba Oil from the Brazilian Amazon." *Genetics and Molecular Biology* 39:248-256. doi: 10.1590/1678-4685-GMB-2015-0098.

Ministério da Saúde. 2015. "Monografia da Espécie *Carapa guianensis* Aubl. (Andiroba)." Brasília: Ministério da Saúde.

Misani, Marta, and De Mey, Albert. 2016. "Managing Complications in Vertical Mammaplasty." *Clinics in Plastic Surgery* 43:359-363. doi: https://doi.org/10.1016/j.cps.2015.12.014.

Morikawa, Toshio, Nagatomo, Akifumi, Kitazawa, Kayako, Muraoka, Osamu, Kikuchi, Takashi, Yamada, Takeshi, Tanaka, Reiko, and Ninomiya, Kiyofumi. 2018. "Collagen Synthesis-Promoting Effects of Andiroba Oil and its Limonoid Constituents in Normal Human Dermal Fibroblasts." *Journal of Oleo Science* 67:1271-1277. doi: 10.5650/jos.ess18143.

Namazi, Mohammad R., Fallahzadeh, Mohammad K., and Schwartz, Robert A. 2011. "Strategies for Prevention of Scars: What Can We Learn from Fetal Skin?." *International Journal of Dermatology* 50:85-93. doi: 10.1111/j.1365-4632.2010.04678.x.

Nayak, Shivananda, Kanhai, Joel, Milne, David M., Pereira, Lexley P., and Swanston, William H. 2011. "Experimental Evaluation of Ethanolic Extract of *Carapa guianensis* L. Leaf for Its Wound Healing Activity Using Three Wound Models." *Evidence-Based Complementary and Alternative Medicine* 2011:419612. doi: 10.1093/ecam/nep160.

Nicolaus, C., Junghanns, S., Hartmann, A., Murillo, R., Ganzera, M., Merfort, I., 2017. *In vitro* studies to evaluate the wound healing properties of Calendula officinalis extracts. *J. Ethnopharmacol.* 196, 94–103. https://doi.org/10.1016/j.jep.2016.12.006.

Ninomiya, Kiyofumi., Miyazawa, Seiya., Ozeki, Kaiten., Matsuo, Natsuko., Muraoka, Osamu, Kikuchi, Takashi, Yamada, Takeshi, Tanaka, Reiko, and Morikawa, Toshio. 2016. "Hepatoprotective Limonoids from Andiroba (*Carapa guianensis*)." *International Journal of Molecular Sciences* 17:591. doi: 10.3390/ijms17040591.

Park, So Y., Lee, Hyun U., Lee, Young-Chul., Kim, Gun H., Park, Edmond C., Han, Seung H., Lee, Jeong G., Choi, Saehae, Heo, Nam S., Kim, Dong L., Huh, Yun S., and Lee, Jouhahn. 2014. "Wound Healing Potential of Antibacterial Microneedles Loaded with Green Tea Extracts." *Materials Science and Engineering C* 42:757-762. doi: 10.1016/j.msec.2014.06.021.

Pereira, Luciano J. B., and Garcia-Rojas, Edwin E. 2015. "Multiple Emulsions: Formation and Application in Microencapsulation of Bioactive Components." *Ciência Rural* 45:155-162. doi: 10.1590/0103-8478cr20140315.

Pieri, Fabio A., Mussi, Maria C., and Moreira, Maria A. S. 2009. "Óleo de Copaíba (*Copaifera* sp.): Histórico, Extração, Aplicações Industriais e Propriedades Medicinais." *Revista Brasileira de Plantas Medicinais* 11:465-472.

Romero, Adriano L. 2007. *Contribuição ao Conhecimento Químico do Óleo-Resina de Copaíba: Configuração Absoluta de Terpenos* [*Contribution to Copaíba Oil-Resin Chemical Knowledge: Absolute Configuration of Terpenes*]. MSc diss., Campinas State University.

Roy, Amit, and Saraf, Shailendra. 2006. "Limonoids: Overview of Significant Bioactive Triterpenes Distributed in Plants Kingdom." *Biological and Pharmaceutical Bulletin* 29:191-201. doi: 10.1248/bpb.29.191.

Sen, Chandan K., Gordillo, Gayle M., Roy, Sashwati., Kirsner, Robert, Lambert, Lynn, Hunt, Thomas K., Gottrup, Finn., Gurtner, Geoffrey C., and Longaker, Michael T. 2009. "Human Skin Wounds: A Major and Snowballing Threat to Public Health and the Economy." *Wound Repair and Regeneration* 17:763-771. doi: 10.1111/j.1524-475X.2009.00543.x.

Shestak, Kenneth C., and Davidson, Edward H. 2016. "Assessing Risk and Avoiding Complications in Breast Reduction." *Clinics in Plastic Surgery* 43:323-331. doi: 10.1016/j.cps.2015.12.007.

Tao, Jing, Rong, Wei, Diao, Xiaoping, and Zhou, Hailong. 2018. "Toxic Responses of Sox2 Gene in the Regeneration of the Earthworm *Eisenia foetida* Exposed to Retnoic Acid." *Comparative Biochemistry and Physiology, Part C* 204:106-112. doi: 10.1016/j.cbpc.2017.12.001.

Taylor, David A. H. 1984. "The Chemistry of the Limonoids from Meliaceae." In *Fortschritte der Chemie organischer Naturstoffe / Progress in the Chemistry of Organic Natural Products*, founded by L. Zeehmeister, edited by W. Hen, H. Grisebaeh, G. W. Kirby, 1-102. Vienna: Springer. doi: 10.1007/978-3-7091-8717-3_1.

Turinetto, Valentina, Vitale, Emanuela, and Giachino, Claudia. 2016. "Senescence in Human Mesenchymal Stem Cells: Functional Changes and Implications in Stem Cell-Based Therapy." *International Journal of Molecular Sciences* 17:1164. doi: 10.3390/ijms17071164.

Veiga Junior, Valdir F., and Pinto, Angelo C. 2002. "The *Copaifera* L. Genus." *Química Nova* 25:273-286. doi: 10.1590/S0100-40422002000200016.

Wanzeler, Ana M. V., Alves Júnior, Sergio M., Gomes, Jessica T., Gouveia, Eduardo H. H., Henriques, Higor Y. B., Chaves, Rosa H., Soares, Bruno M., Salgado, Hugo L. C., Santos, Alberdan S., and Tuji, Fabrício M. 2018. "Therapeutic Effect of Andiroba Oil (*Carapa guianensis* Aubl.) Against Oral Mucositis: An Experimental Study in Golden Syrian Hamsters." *Clinical Oral Investigations* 22:2069-2079. doi: 10.1007/s00784-017-2300-2.

Wen, Kuo C., Lin, Shiuan P., Yu, Chung P., and Chiang, Hsiu M. 2010. "Comparison of Puerariae Radix and its Hydrolysate on Stimulation of Hyaluronic Acid Production in NHEK Cells." *The American Journal of Chinese Medicine* 38:143-155. doi: 10.1142/S0192415X10007725.

Yang, Yuwei, Hu, Haicong, Wang, Wenqi, Duan, Xiaojie, Luo, Shilin, Wang, Xiongfei, and Sun, Yikun. 2017. "The Identification of Functional Proteins from Amputated Lumbricus *Eisenia fetida* on the

Wound Healing Process." *Biomedicine & Pharmacotherapy* 95:1469-1478. doi: 10.1016/j.biopha.2017.09.049.

Yang, Yuwei, Sun, Yujie, Zhang, Na, Li, Jianhao, Zhang, Chenning, Duan, Xiaojie, Ding, Yuting, Zhao, Renyun, Zheng, Zhuhong, Geng, Di, and Sun, Yikun. 2019. "The Up-Regulation of Two Identified Wound Healing Specific Proteins-HSP70 and Lysozyme in Regenerated *Eisenia fetida* Through Transcriptome Analysis." *Journal of Ethnopharmacology* 237:64-73. doi: 10.1016/j.jep.2019.03.047.

Zhang, Xue, Xu, Jie K., Wang, Jue, Wang, Nai L., Kurihara, Hiroshi, Kitanaka, Sumumu, and Yao, Xin S. 2007. "Bioactive Bibenzyl Derivatives and Fluorenones from *Dendrobium nobile*." *Journal of Natural Products* 70:24-28. doi: 10.1021/np060449r.

In: A Closer Look at Fibroblasts
Editor: Justin O'Shane

ISBN: 978-1-53616-977-5
© 2020 Nova Science Publishers, Inc.

Chapter 4

LASER THERAPY ASSOCIATED WITH GUARANÁ AS A THERAPEUTIC ALTERNATIVE ON THE SKIN OF OXY-INFLAMMATORY METABOLISM

Daíse Raquel Maldaner[1], Cibele Ferreira Teixeira[2], Marta Maria Medeiros Frescura Duarte[1], Verônica Farina Azzolin[2], Ivana Beatrice Mânica da Cruz[2], Ednea Aguiar Ribeiro[3], Neida Luiza Pellenz[4] and Fernanda Barbisan[2]

[1]Universidade Luterana do Brasil - Campus Santa Maria, Santa Maria, RS, Brazil
[2]Biogenomic Laboratory, Department of Morphology, Universidade Federal de Santa Maria, Santa Maria, RS, Brazil
[3]Fundação Universidade Aberta da Terceira Idade, Manaus, AM, Brazil
[4]Universidade Federal de Santa Maria, Canpus Palmeira das Missões, Brazil

1. Physiopathology of Skin Aging

Skin aging is induced by genetic factors (chronological aging or intrinsic aging) and environmental or external factors, especially represented by sun exposure (photoaging, extrinsic or actinic aging). It is known that the cellular and molecular mechanisms of these factors are the same, that is, a superposition of the biological effects of extrinsic stressors occurs on intrinsic aging (Luca et al., 2013).

Therefore, it is emphasized that the organism as a whole undergoes a natural chronological aging process, simultaneously affecting all organs and tissues, because the cells possess a finite multiplication capacity, leading to cell senescence (Dihl and Lehmann, 2011; Kamizato, Brito, 2014).

This intrinsic aging is slow and gradual, and is directly proportional to the period already lived, characterized by low cellular regeneration and compromised immune responses, inducing successive damage of cellular structures. In addition, there are modifications in the genetic transcription of innumerable DNA molecules, proteins and enzymes, rendering the tissues inefficient in their functions (Bhatia-Dey et al., 2016).

The genetically programmed intrinsic aging is called the biological clock and is directly related to telomeres. These structures are constituted by rows of proteins and DNA, and act as protectors for the chromosomes, preventing recombination and fusion of the final sequences with other chromosomes, aid in the recognition of DNA damage, participate in the regulation of gene expression, ensuring that the relevant genetic information is perfectly copied when the cell doubles. With each cell division, the telomeres undergo shortening and, with this, the replicative capacity of the cells decreases until generating cellular senescence and ending in programmed cell death or apoptosis. The control of this process is the responsibility of a ribonucleicoproteic enzyme complex called telomerase, capable of synthesizing telomeric DNA replications and recovering the cell division capacity, when necessary (Borges and Scorza, 2016).

This mechanism of senescence directly affects the components of the dermal matrix, as it acts on the expression of fibroblasts, which remain in stationary phase for a long period of time in the dermis. These will only proliferate when there is stimulation or telomere shortening (Aldag et al., 2016; Harris, 2009).

Cells in senescence lose the ability of cell division, and it occurs a reduction or absence of telomerase activity and antioxidant capacity. In addition, there is a higher production of reactive oxygen species (ROS), due to mitochondrial dysfunction. At the end of the senescence period, cell death occurs by apoptosis (Höhn et al., 2017; Lake, 2016).

Intrinsically aged skin is atrophic, the dermo-epidermal junction is flattened and metabolic activity is slow. There is an increase in the size of the keratinocytes, loss of the elastic capacity by reduction in the activity of the fibroblasts, resulting in the increase of the expression lines. It is a thin, pale skin with fine wrinkles, dry, and may present benign neoplasms. Between the third and seventh decade of life, the rate of epidermal cell turnover decreases by 50% and the fibroblast growth rate by around 30 to 50% (Borges, Scorza, 2016; Harris, 2009; Höhn et al., 2017)

In the skin, the manifestations of aging are larger and visibly detected due to actinic damage generated by contact with the external environment. This damage is triggered by a photochemical process between cutaneous chromophores and ultraviolet radiation (UVR) with ROS release, and a signaling cascade of inflammatory cytokines, increased expression of cell surface receptors and transcriptors, changes in the DNA molecule resulting in telomere shortening and release of enzymes responsible for the degradation of the structures that support the skin. Cells with damaged membranes are subject to apoptosis and impairment in active substance transport, inducing damage to biological systems (Lago, 2016; Rapisarda et al., 2017).

Daily environmental exposure to stressors, including UVR and fumes increases oxidative stress. This results in tissue damage due to antioxidant depletion associated with increased ROS production. Multiple biochemical pathways that are triggered by ROS overload result in activation of activating protein 1 (AP-1) and consequent suppression of receptor-transforming growth factor II (TGF-β-R2), overexpression of matrix degradation metalloproteinases (MMPs) which are collagenases, and increased inflammation through nuclear factor kappa B (NF-kB). In addition, UVR also causes direct damage to structural skin proteins (Bhatia-Dey et al., 2016; Fabi; Sundaram, 2014; Souza, 2015).

The deregulation of TGF-β signaling plays a prominent role in the pathogenesis associated with extracellular matrix. For example, up-regulation of TGF-β signaling causes abnormal protein accumulation in affected tissues. In contrast, down-regulation of TGF-β signaling negatively regulates collagen homeostasis and has significant impact on connective tissue aging resulting in thinning of aged skin (Quan; Fisher, 2015; Xu et al., 2018).

Advancing of age and the accumulation of extrinsic damage promotes the decline of skin functions, including a reduction in its ability to act as a protective barrier against the environment and providing drier skin (xerosis). Interruption of the epidermal barrier function elevates the levels of proinflammatory cytokines in the skin (Velarde, 2017).

Skin aged by extrinsic factors, in contrast to intrinsically aged skin, is characterized by pigmentation changes (age spots), deep wrinkles, epidermal atrophy with corneocyte accumulation, resulting in a rough, dull or leathery appearance. In addition, elastosis occurs, characterized by the accumulation of amorphous elastin with reduced elasticity, collagen fibers become thicker, fragmented and more soluble. There are still changes in subcutaneous tissue with reduced vascularization and structural changes that also compromise the functions and appearance of the skin (Aldag et al., 2016).

Most of the factors that cause premature skin aging are due to the action of oxidative stress generated by overexposure to environmental factors (Cole et al., 2018; Lago, 2016; Velarde, 2017).

1.1. Role of Inflammatory Oxidative Metabolism in Skin Aging

Cell aging can be explained by ROS theory, which is based on aging induced by deleterious effects on cell organelles caused by ROS, produced mainly in mitochondria, where aerobic metabolism occurs (Andrisic et al., 2018; Harris, 2009; Teixeira; Guariento, 2010).

Oxidative aerobic metabolism is a biochemical process in which oxygen is used for energy production (Perl, 2013). During oxidative metabolism, ROS generation occurs in a continuous and physiological process, with relevant biological functions such as the participation of egg fertilization and the body's defense mechanisms, but excessive production can lead to oxidative damage (Barbosa et al. 2010).

Inside mitochondria oxygen passes through the electron transport chain generating adenosine triphosphate (ATP), that is metabolic energy. However, oxygen reduction is incomplete and leads to the generation of different ROS, such as superoxide radical (O_2^-) and hydrogen peroxide (H_2O_2), considered the main contributor to oxidative damage (Vitale et al., 2013). H_2O_2 is capable of giving rise to other ROSs, especially the hydroxyl radical (OH^-), which, having a very short half-life, can hardly be sequestered *in vivo*. Also about OH^-, it has a high affinity for DNA, causing damage to it (Andrisic et al., 2018; Barbisan, 2014; Perl, 2013).

H_2O_2 is extremely unstable and can permeate through lipid membranes. Additionally, H_2O_2 can also be converted to OH^- in the presence of iron (Fe^{2+}) or copper (Cu^{2+}) ions through the Fenton reaction. The transition metal ions, iron and copper, also catalyze the reaction between O_2^- and H_2O_2, generating OH^- in the Haber-Weis reaction (Barandalla et al., 2017; Chen et al., 2012).

The OH^- radical is one of the most unstable ROSs in biological systems because it is capable of reacting with various biomolecules such as DNA, enzymes, proteins and amino acids. It can also react with membrane phospholipids, promoting their disorganization, forming alkoxyl radicals (LO^-), peroxyl radical (LOO^-), and lipid hydroperoxides (LOOH), precursors of lipid peroxidation (Cadet; Wagner, 2016; Souza, 2015). Lipoperoxidation termination occurs when the radicals produced propagate until they destroy themselves leading to rupture of their structure, decay in metabolite exchange and, in a condition of excessive oxidative damage, cellular apoptosis occurs. Lipid peroxidation can be measured by the 2-thiobarbituric acid test (TBARS) (Engers et al., 2011; Furukawa et al., 2004).

Damage to protein structures leads to fragmentation of polypeptide chains, formation of protein-protein bonds, and modifications to amino acids in side chains. As a consequence of these reactions, there can occur losses in enzymatic activity, difficulties in the active transport process, cytolysis and cell death. Protein carbonylation is a biomarker for assessing oxidative damage to proteins caused by ROS. Refers to irreversible, non-enzymatic protein modification (Sies et al., 2017; Veskoukis et al., 2012).

It should be noted here that ROS are fundamental for organism survival and that basal levels of reactive species are important for the maintenance of organism homeostasis, since ROS such as H_2O_2 and nitric oxide (ON) act as signals for cell proliferation, migration, survival and differentiation (Machado, 2014). However, high levels of ROS can cause damage to cells, tissues and so disrupt the body's homeostasis. Thus, humans have 2 main antioxidant defense systems: exogenous and endogenous. The exogenous system is acquired via food through the consumption of bioactive molecules, which contribute to the better functioning of the endogenous system, which is composed of antioxidant enzymes such as Superoxide Dismutase (SOD), Catalase (CAT) and Glutathione Peroxidase (GPX) (Kurutas, 2016; Montagner, 2010).

Enzymatic antioxidants are the main form of defense against ROS, as they are intended to catalyze the reactions to neutralize them, produced during the metabolism of aerobic respiration and substrate oxidation. The cell's first line of defense against excess ROS has the detoxifying action of oxidizing agents before they cause injury. It consists of SOD, which catalyzes the reaction of O_2^- in H_2O_2, in turn CAT, reduced glutathione and GPX catalyze the degradation of H_2O_2 in H_2O (water). These enzymes are able to neutralize oxidizing agents and maintain them at appropriate levels in the body (Neves et al., 2014).

When the generation of ROS increases and/or the antioxidant defenses decrease, the oxidative stress sets in. Acute exposure to UVR, for example, eliminates CAT activity in the skin and increases protein oxidation. Among all environmental factors, it is estimated that UVR contributes up to 80% and is the most important environmental factor in the development of skin cancer and skin aging. The primary mechanism by which UVR initiates molecular responses in human skin is via chromophore activation and photochemical generation of ROS, especially O_2^-, H_2O_2, OH^- and singlet oxygen ($^1O_2$) formation. UVR penetrates the skin, reaches cells and is absorbed by DNA, leading to the formation of photoproducts that inactivate DNA functions (Poljšak; Dahmane, 2012).

Studies show that oxidative stress is closely associated with the aging process and cell death by apoptosis. ROS has been considered a prerequisite for the inflammatory and apoptotic process. Thus, oxidative stress would play a central role in the process of cell aging, as well as pathologies and loss of organism homeostasis (Brand et al., 2017; Chandra et al., 2000; Sinha et al., 2013).

The inflammatory process is a reaction of the organism to an infection or any tissue damage. Since the skin is constantly exposed to external aggressors such as UVR, pollution, among others, it suffers from successive inflammation-inducing aggressions. In the formation of this process, initially there is a reflex vasoconstriction followed by vasodilation, erythema, heat and pain (McMorrow; Murphy, 2011). Vasodilation induces increased vascular permeability, allowing fluid to infiltrate local tissues, promoting edema formation, increasing skin hypersensitivity and tension and causing the release of chemical mediators to the site (Han et al., 2017; Zimmermann et al., 2011).

Inflammatory mediators such as pro-inflammatory interleukin 1 (IL-1) cytokines, tumor necrosis factor alpha (TNF-α) and interferon gamma (IFN-γ) and anti-inflammatory cytokines such as IL-10 are typically produced by macrophages and also by epithelial, endothelial or fibroblast cells. TNF-α and IL-1 will attract more neutrophils to the site as well as activate them. Lymphocytes promote macrophage activation, leading them to secrete proteases and cytokines such as IL-6 and IFN-γ, which will activate other cells, thus forming a cycle of cell recruitment and activation (Carneiro et al., 2011; Czemplik et al., 2017; Guerreiro et al., 2011).

TNF-α acts in different parts of the body and is considered one of the most important cytokines related to inflammatory and immune processes. The main physiological effect of TNF-α is to produce and activate the immune and inflammatory response by recruiting neutrophils and monocytes to the site of infection, and activating them (André; Tofalini, 2008; Rambo, 2013). When released in small concentrations, TNF-α acts on endothelial cells, generating vasodilation and stimulating them to secrete a group of chemotaxically acting cytokines relative to leukocytes, thus producing a local inflammatory process that allows the combat of infectious conditions (Han et al., 2017; Rambo, 2013).

IFN-γ is a cytokine that has cell-mediated effect of the immune system, and with important immunosuppressive property, regulating the cell-mediated immune response (Portella, 2011).

IL-6 is a cytokine acting on both innate and adaptive immune responses. This interleukin stimulates hepatocytes to produce messenger ribonucleic acid (mRNA) for production of acute phase proteins. It is synthesized by monocytes, fibroblasts, endothelial cells, cells in response to microorganisms and stimulation by other cytokines, mainly IL-1 and TNF-α (Souza et al., 2008).

Cytokines act mainly in two ways: by stimulating NF-kB levels in the cytoplasm or by increasing levels of mitogen-activated protein kinases (MAP kinases) in cells. Activation of the MAP kinase family in the skin results in increased transcription of the c-jun transcription factor, which together with the constitutively expressed transcription factor c-Fos form the AP-1 complex. AP-1 is required for activation of MMPs, which are responsible for the degradation of collagen and elastic dermis fibers (Han et al., 2017; Souza, 2015).

Therefore, neutralization and removal of agents that cause inflammation and/or oxidative stress assist the immune system in a positive response, capable of resolving the process, otherwise, as a response to the damage of cellular DNA suffered, various cellular responses are induced, such as the transcriptional activation of the p53 gene (tumor suppressor protein), which directly affects the cell cycle and induces the apoptotic cascade (Cadet; Wagner, 2016).

1.2. Performance of Growth Factors in Maintaining Skin Integrity

In the tissue repair process, growth factors play a fundamental role in stimulating and activating cell proliferation, angiogenesis, mitogenesis and genetic transcription. Growth factors act to inhibit or stimulate gene expression of injured target cells. By transmitting modulatory signals, they regulate stimulating and inhibiting growth processes, such as proliferation, differentiation, migration, and adherence. They also promote cell chemotaxis, being able to induce the migration of several cells, besides stimulating angiogenesis and extracellular matrix synthesis (Carvalho, 2016).

The major growth factors involved in tissue repair are platelet-derived growth factor (PDGF), transforming growth factor alpha (TGF-α), epidermal growth factor (EGF), vascular endothelial growth factor (VEGF), fibroblast growth factor (FGF) and keratinocyte growth factor (KGF) (Borges; Scorza, 2016; Carvalho, 2016; Thomson et al., 2019).

FGFs are secreted by macrophages, fibroblasts, mast cells and endothelial cells (Szwed; Santos, 2016). FGF-1 is involved with osteoblast proliferation and differentiation and osteoclast inhibition, promotes cell migration and angiogenesis, and is a mitotic agent for keratinocytes, fibroblasts and vascular endothelial cells. FGF-2 is involved with the growth of fibroblasts, myoblasts, endothelial cells and keratinocytes, increased fibronectin production, angiogenesis stimulation, endothelial cell proliferation, collagen synthesis, matrix synthesis and wound retraction (Carvalho, 2016). Fabi; Sundaram, 2014).

KGF is a highly specific growth factor that exerts mitogenic effects on the cellular epithelium and has been reported as a key factor in the healing process as it protects epithelial cells from damage and plays an important role in repairing damaged cells (Yang et al., 2018; Yu et al., 2017).

2. Low Potency Laser

Laser (Light Amplification by Stimulated Emission of Radiation) is a type of light energy that has precise characteristics and properties such as monochromaticity, coherence and polarization, and can be divided, according to its wavelength, into high, medium or low power laser for use in different pathologies (Agne, 2013; Belykh et al., 2017).

The difference between the various types of lasers is given by the wavelength, and the longer the wavelength, the greater its penetration power, which can be emitted continuously or pulsed. The wavelength of the low power laser (LBP), capable of providing a significant result in the regulation of cellular functions, is between 620 and 860nm, corresponding to the emission of red and infrared light, therefore the LBP is the most indicated for skin treatments due to their direct interaction with epidermal and dermal cells including fibroblasts, macrophages and endothelial cells (Rambo, 2013).

Its power is expressed in watts (W), ranging from deciwatts to megawatts and the energy or dose is measured in joules per square centimeter (J/cm2), being equal to the power multiplied by the application time. The effects generated when this energy is deposited in tissues are called photobiomodulation (Agne, 2013; Lins et al., 2010).

Photobiomodulation begins when a suitable molecule, known as a chromophore, absorbs a photon of light energy of a suitable wavelength and an electron enters a state of excitation. A chromophore is a molecule that gives a color to a compound and includes hemoglobin, oxyhemoglobin, myoglobin, cytochrome, flavin, flavoproteins, porphyrins, and melanin. In many cases, the target chromophore is the iron and copper-containing cytochrome C oxidase (CCO) enzyme located in the mitochondrial respiratory chain. OCC excitation results in increased production of ATP, NADH, RNA and increased cellular respiration. Other effects include increased CAT and SOD antioxidant enzymes, increased NO release, and increased ROS production (Nestor et al., 2017; Pinto, 2011; Silveira et al., 2009).

However, it is noteworthy that, depending on the dose, wavelength, exposure time and tissue type, laser therapy may interfere in different ways with antioxidant defense and tissue repair mechanisms (Belykh et al., 2017; Silva et al., 2016).

It is noteworthy that the effects of LBP irradiation are dependent on the molecular absorption of its energy and its transformation into certain biological processes. Upon contact with the tissue, the energy deposited in the tissue is absorbed through the chromophores, resulting in primary (direct) actions that trigger secondary (indirect) actions. Direct actions include the local effects on ATP, DNA and protein synthesis, membrane normalization and functional restoration, as mentioned above; while indirect actions are related to systemic effects such as microcirculatory stimulation and anti-inflammatory effects by reducing proinflammatory cytokines such as TNF-α, IL-2 and IL-6 (Agne, 2013; Lins et al., 2010; Piva et al., 2011; Silva et al., 2016).

Side effects result from amplification of the light energy response and transmission of this response to other parts of cells, resulting in physiological effects such as changes in cell membrane permeability, increased cell metabolism, DNA and RNA synthesis, increased fibroblast proliferation, activation of T lymphocytes, macrophages and mast cells, increased endorphin synthesis and bradykinin decrease (Chaves et al., 2014; Yun; Kwok, 2017).

However, in order to obtain a good therapeutic result, a correct diagnosis of the pathology is necessary, together with the choice of the appropriate resource. In LBP laser therapy, it is important to know the energy density to be deposited, the correct way to deposit it, the desired photobiological reaction, the pulse rate and the number and frequency of applications. The energy density (dose) that is deposited on the treated area for biostimulation purposes, according to the majority of studies already performed, should remain between 1 J/cm^2 and 6 J/cm^2 (Agne, 2013; Pereira, 2013; Szezerbaty et al., 2018). For Piva et al. (2011), the indicated therapeutic dosages may vary from 1.8 to 16 J/cm^2, those being the lowest doses, up to 8 $J/cm2$ indicated for anti-inflammatory and healing action and the highest doses for inhibitory effects.

The study by Ginani et al. (2018) investigated the effects of LBP on stem cell proliferation and viability of human teeth, using as laser parameters 660 nm, 30 mW, comparing doses of 0.5 J/cm² and 1.0 J/cm². The study concluded that the 1.0 J/cm² dose was more effective in promoting dental stem cell proliferation and maintaining cell viability when compared with control and lower dose.

Phototherapy with the correct dosage stimulates IL-6 expression, cell proliferation and cell migration in diabetic wound cells studied *in vitro*. The dose of 5 J/cm² stimulates the healing of diabetic wounds, while 16 J/cm2 is inhibitory and damage-inducing (Houreld; Abrahamse, 2007).

5 J/cm² LBP (660 nm) modulates cell viability, regulates VEGF expression, and decreases IL-6 and mRNA expression in L929 fibroblast cell culture in an *in vitro* study (Szezerbaty et al., 2018).

Another recent study evaluated the effect of 660 nm LBP irradiation on submandibular glands of diabetic rat models and found a significant reduction in the expression of apoptosis biomarkers (caspase 3), as well as a decrease in p53 protein levels, which suggests an action of LBP in the control of diabetes-induced apoptosis (Fukuoka et al., 2017).

It is also noteworthy that the most documented effects of LBP are scar tissue remodeling, decreased inflammatory process, stimulation of collagen synthesis and epithelialization, repair of fractures and skin wounds in studies developed in different *in vitro* and *in vitro* models (Borges; Scorza, 2016; Pinto, 2011; Ramos et al., 2018). Importantly, the application of LBP is absolutely contraindicated in the region of the eyeball, in neoplastic tissues and the pregnant uterus; and relatively contraindicated in cases of bleeding, neuropathies, infections and gonads (Christofoletti et al., 2010).

Studies are being conducted associating LBP with natural products that induce better tissue regeneration, in order to support and accelerate treatment and thus amplify cellular responses. As an example, the study by Catarino et al. (2015) found that the application of LBP (670 nm, dose of 4.93 J/cm²) alone or combined with *Solidago chilensis* extract (popularly known as Brazilian arnica) promoted favorable responses in the tissue repair of second degree burns in an experimental animal model. And, more recently, an *in vitro* study by Carvalho et al. (2018) demonstrated that the association of

LBP (660 nm and 3 J/cm² dose) with *Aloe vera* favorably affected the cell viability of dental pulp fibroblasts and microbial control, appearing as an alternative therapy for intracanal endodontic treatment.

3. Guaraná (*Paullinia Cupana* Var. Sorbilis Mart)

Guarana is a plant belonging to the Sapindaceae family, native to the Amazon basin, commonly used in traditional medicine. In nature, it grows like a woody vine that can reach 10 meters in length, but when indoors, it grows like a bush reaching about two to three meters high. It has a characteristic red fruit that, when ripe, partially opens, displaying up to three seeds. Roasted seed extracts have been used in medicines and beverages since pre-Columbian times as stimulants, aphrodisiacs and toners (Krewer et al., 2011; Machado et al., 2015; Peixoto et al., 2017).

Guarana is a Brazilian diet food with high phytotherapeutic potential, mainly because it is rich in xanthines and catechins. This statement is based on previous research by Angelo et al. (2008), who investigated the guarana transcriptome and identified a large number of transcripts of bioactive compounds present in this plant, which are similar to those found in *Camelia sinensis* (green tea).

The chemical constitution of guarana is quite complex and among the components are alkaloids such as theophylline, caffeine, theobromine, as well as terpenes, flavonoids, amides, saponins, fats, starch, choline and pigments. Bittencourt et al. (2013) analyzed the chemical composition of guarana from Maués-Amazonas via high performance liquid chromatography (HPLC) and the main bioactive compounds found were caffeine = 12,240 mg/g, theobromine = 6,733 mg/g and catechins = 4,336 mg/g.

It is believed that the chemical composition quite rich in compounds with recognized health-beneficial activities holds the key to several scientifically proven guaraná pharmacological actions, such as: antioxidant activity (Krewer et al., 2011; Machado et al., 2015), antiplatelet, anti-inflammatory (Krewer et al., 2014), energetic (Campos et al., 2011), thermogenic (Suleiman et al., 2016), antiobesogenic, hypolipid, in neurocognitive modulation (Dalonso; Petkowicz, 2012), antibacterial and antifungal (Basile et al., 2005; Yamaguchi-Sasaki et al., 2007), antitumor (Hertz et al., 2015), reduced levels of oxidized LDL and cardiovascular disease (Portella et al., 2013), cytoprotective (Kober et al., 2016), long-term antidepressant effect (Otobone et al., 2007), protective effect against methylmercury toxicity (Arantes et al., 2016).

In addition, the study by Machado et al. (2015) analyzed the action of guarana supplementation on senescence of mesenchymal adipose cells (ASC), obtained from human liposuction, and whether this action would involve the differential regulation of cellular oxidation metabolism. Senescent ASC cells received a 5 mg/mL guarana supplement in the eighth passage, i.e., when they lost approximately 25% of their proliferative capacity. These cells showed an improvement in their proliferation as well as decreased oxidative stress markers after treatment.

The antioxidant effects of guarana (Paullinia cupana var. Sorbilis Mart.) Extract on ON and other compounds generated from the degradation of sodium nitroprusside (SNP) in embryonic fibroblast culture (NIH-3T3 cells) were evaluated in a study of Bittencourt et al. (2013), verifying that guarana bioactive compounds reversed SNP toxicity, especially at lower concentrations (< 5 mg), which decreased cell mortality, lipid peroxidation, DNA damage and cellular oxidation, as well as increased SOD levels.

Also, when tested in the *in vivo* Caenorhabditis elegans model, a nematode, the roasted extract of roasted guarana seeds improved resistance against oxidative stress, increased shelf life and attenuated aging markers such as decline in muscle activity related to age (Peixoto et al., 2017).

4. ASSOCIATION BETWEEN GUARANÁ AND LOW POWER LASER. EXPERIENCE OF NA *IN VIVO* STUDY

Maldaner et al. (2019) conducted an *in vitro* study using a commercial human dermal fibroblast (HFF-1) strain to evaluate the antioxidant, anti-inflammatory, anti-apoptotic, and proliferative effects of the association of guarana extract with the low power laser.

Initially, cells were cultured under appropriate conditions and treated with different concentrations of guarana hydroalcoholic extract (1, 3, 5, 10 and 30 µg/mL) for 72 hours. Next, inflammatory, oxidative and apoptotic markers were evaluated, and the concentration that obtained the best result against these parameters was chosen to be used in subsequent tests in association with low power laser therapy.

The fibroblasts were then treated with guarana extract and, after 2 hours, were irradiated with a 660nm laser, with 35mW output power and 16 Hz frequency in point continuous wave mode. The dose administered was 4 J/cm^2, with exposure time of 14s. The irradiation process was achieved at room temperature (18-25°C) and all treatments were performed with a distance of 35 mm between the laser and the irradiated area.

After 72 hours of irradiation, the following parameters were analyzed: (1) markers of oxidative metabolism: ROS levels, protein carbonylation, lipoperoxidation, DNA oxidation and antioxidant enzyme levels; (2) inflammatory markers: gene expression (after 24h treatment) and protein levels (after 72h treatment) of IL-1β, IL-6, TNF-α and IL-10 cytokines (3) apoptotic markers: gene expression (after 24h treatment) and protein levels (after 72h treatment) of caspases 1, 3 and 8; (4) cell proliferation analysis: cell cycle analysis by flow cytometry and quantification of the levels of two cell signaling growth factors (FGF-1 and KGF).

As a result, in the primary analysis of the effect of different guarana concentrations on inflammatory, oxidative and apoptotic parameters, it was observed that the concentration of 5 µg/mL did not induce the production of proinflammatory cytokines, significantly increasing anti cytokine levels. IL-10 did not increase caspase levels or DNA oxidation in relation to the control group, which is the concentration of choice for later analysis in combination with low-level laser therapy.

On oxidative markers, it was observed that the combined treatment of guarana (5 µg/mL) with low-level laser therapy (4 J/cm2) reduced protein carbonylation and DNA oxidation levels, but increased levels of ROS and lipoperoxidation compared to the untreated control group. In addition, levels of antioxidant enzymes SOD, CAT, and GPX significantly increased in the guarana plus laser therapy-treated cell group compared with the untreated group.

The combination of guarana treatment with low-level laser was able to decrease protein levels and gene expression of IL-1β, IL-6, TNF-α proinflammatory cytokines, and to increase protein levels and gene expression. anti-inflammatory cytokine IL-10 compared to the control group.

Protein and gene levels of caspases 1, 3 and 8, which are involved in the apoptotic route, were decreased with the associated treatment of guarana with the laser when compared with the control group.

In the evaluation of fibroblast cell proliferation through flow cytometry analysis of the cell cycle, it was shown that the combined effect of guarana with the low power laser significantly increased the number of cells in the S phase of the cell cycle compared to the group control. The combination of treatments was also able to increase the levels of FGF-1 and KGF cell signaling growth factors in relation to the control.

Importantly, many of the reported effects were also observed in treatment with guarana or low-level laser alone and are demonstrated in the article.

In conclusion, despite methodological limitations related to *in vitro* studies, the results of this work may be clinically relevant, since the isolated and associated treatment of guarana and low-level laser seems to promote fibroblast biostimulation from antioxidant effects, anti-inflammatory, anti-apoptotic and proliferative in these cells.

References

Agne, J. E. (2013). *Electrotermophototherapy*. 2 ed. Santa Maria.

Aldag, C., Teixeira, D. N. & Leventhal, P. S. (2016). Skin rejuvenation using cosmetic products containing growth factors, cytokines, and matrikines: a review of the literature. *Clin Cosmet Investig Dermatol, 9,* 411-419.

André, A. L. R. & Tofalini, F. S. (2008). *What is the role of the kappa B nuclear factor (NF-κB) in obesity?* Completion of course work (Graduation in Nutrition) - University Center of Belo Horizonte, Belo Horizonte, MG, Brazil.

Andrisic, L., Dudzik, D., Barbas, C., Milkovic, L., Grune, T. & Zarkovic, N. (2018). Short overview on metabolomics approach to study pathophysiology of oxidative stress in cancer. *Redox Biol*, *14*, 47-58.

Angelo, P. C., Nunes-Silva, C. G., Brígido, M. M., Azevedo, J. S., Assunção, E. N., Sousa, A. R., Patrício, F. J., Rego, M. M., Peixoto, J. C., Oliveira, W. P. Jr., Freitas, D. V., Almeida, E. R., Viana, A. M., Souza, A. F., Andrade, E. V., et al. (2008). Guarana (Paullinia cupana var. sorbilis), an anciently consumed stimulant from the Amazon rain forest: the seeded-fruit transcriptome. *Plant Cell Rep*, *27*, 117-124.

Arantes, L. P., Peres, T. V., Chen, P., Caito, S. W., Aschner, M. & Soares, F. A. A. (2016). Guarana (Paullinia cupana Mart.) attenuates methylmercury-induced toxicity in Caenorhabditis elegans. *Toxicol Res*, *5* (VI), 1629-1638.

Barandalla, M., Shi, H., Xiao, H., Colleoni, S., Galli, C., Lio, P., Trotter, M. & Lazzari, G. (2017). Global gene expression profiling and senescence biomarker analysis of hESC exposed to H_2O_2 induced non-cytotoxic oxidative stress. *Stem Cell Res Ther*, *8* (I), 160.

Barbisan, F. (2014). *Pharmacogenetic and pharmacogenomic effect of methotrexate on the cytotoxicity of peripheral blood mononuclear cells*. Dissertation (Master's degree in Pharmacology) - Federal University of Santa Maria, Santa Maria, RS, Brazil.

Barbosa, K. B. F., Costa, N. M. B., Alfenas, R. C. G., de Paula, S. O., Minim, V. P. R. & Bressan, J. (2010). Oxidative stress: concept, implications and modulating factors. *Brazilian Journal of Nutrition*, *23* (IV), 629-643.

Basile, A., Ferrara, L., Pezzo, M. D., Mele, G., Sorbo, S., Bassi, P. & Montesano, D. (2005). Antibacterial and antioxidant activities of ethanol extract from Paullinia cupana Mart. *J Ethnopharmacol*, *102* (I), 32-36.

Belykh, E., Yagmurlu, K., Martirosyan, N. L., Lei, T., Izadyyazdanabadi, M., Malik, K. M., Byvaltsev, V. A., Nakaji, P. & Preul, M. C. (2017). Laser application in neurosurgery. *Surg Neurol Int, 8*, 274.

Bento, B. S. (2015). *Skin photoaging: process and products.* Dissertation (Integrated master's degree in Pharmaceutical Sciences) - Higher Institute of Health Sciences Egas Moniz, Almada, Portugal.

Bhatia-Dey, N., Kanherkar, R. R., Stair, S. E., Makarev, E. O. & Csoka, A. B. (2016). Cellular senescence as the causal nexus of aging. *Front Genet, 7*, 13.

Bittencourt, L. S., Machado, D. C., Machado, M. M., dos Santos, G. F., Algarve, T. D., Marinowic, D. R., Ribeiro, E. E., Soares, F. A., Barbisan, F., Athayde, M. L. & Cruz, I. B. (2013). The protective effects of guaraná extract (Paullinia cupana) on fibroblast NIH-3T3 cells exposed to sodium nitroprusside. *Food Chem Toxicol, 53*, 119-125.

Borges, F. S. & Scorza, F. A. (2016). *Therapeutics in aesthetics: concepts and techniques.* 1 ed. Sao Paulo: Phorte.

Brand, R. M., Epperly, M. W., Stottlemyer, J. M., Skoda, E. M., Gao, X., Li, S., Huq, S., Wipf, P., Kagan, V. E., Greenberger, J. S. & Falo, L. D. Jr. (2017). A topical mitochondria-targeted redox cycling nitroxide mitigates oxidative stress induced skin damage. *J Invest Dermatol, 137* (III), 576-586.

Cadet, J. & Wagner, J. R. (2016). Radiation-induced damage to cellular DNA: Chemical nature and mechanisms of lesion formation. *Radiat Phys Chem, 128*, 54-59.

Campos, M. P., Hassan, B. J., Riechelmann, R. & Del Giglio, A. (2011). Cancer-related fatigue: a review. *J Braz Med Assoc, 57* (II), 206-214.

Carneiro, J. R., Fuzii, H. T., Kayser, C., Alberto, F. L., Soares, F. A., Sato, E. I. & Andrade, L. E. (2011). IL-2, IL-5, TNF-α and IFN-γ mRNA expression in epidermal keratinocytes of systemic lupus erythematosus skin lesions. *Clinics, 66* (I), 77-82.

Carvalho, M. R. (2016). *Growth factors for treatment of venous ulcers: systematic review and meta-analysis. Dissertation* (Master's degree in Science in Health Care) - Fluminense Federal University, Niterói, RJ, Brazil.

Carvalho, N. C., Guedes, S. A. G., Albuquerque-Júnior, R. L. C., de Albuquerque, D. S., Araújo, A. A. S., Paranhos, L. R., Camargo, S. E. A. & Ribeiro, M. A. G. (2018). Analysis of Aloe vera cytotoxicity and genotoxicity associated with endodontic medication and laser photobiomodulation. *J Photochem Photobiol B, 178*, 348-354.

Catarino, H. R., de Godoy, N. P., Scharlack, N. K., Neves, L. M., de Gaspi, F. O., Esquisatto, M. A., do Amaral, M. E., Mendonça, F. A. & dos Santos, G. M. (2015). InGaP 670-nm laser therapy combined with a hydroalcoholic extract of Solidago chilensis Meyen in burn injuries. *Lasers Med Sci, 30* (III), 1069-1079.

Chandra, J., Samali, A. & Orrenius, S. (2000). Triggering and modulation of apoptosis by oxidative stress. *Free Radic Biol Med, 29* (III-IV), 323-333.

Chaves, M. E. A., Araújo, A. R., Piancastelli, A. C. C. & Pinotti, M. (2014). Effects of low-power light therapy on wound healing: LASER x LED. *Braz Annals Dermatol, 89* (IV), 616-623.

Chen, L., Hu, J. Y. & Wang, S. Q. (2012). The role of antioxidants in photoprotetion: a critical review. *J Am Acad Dermatol, 67* (V), 1013-1024.

Christofoletti, D. C., Souza, M. V. G., Chingui, L. J. & Severi, M. T. M. (2010). Laser and microcurrent actions in cutaneous lesions. *Student Scientific Initiation Production Yearbook, 13* (XVI): 93-101.

Cole, M. A., Quan, T., Voorhees, J. J. & Fisher, G. J. (2018). Extracellular matrix regulation of fibroblast function: redefining our perspective on skin aging. *J Cell Commun Signal, 12* (I), 35-43.

Czemplik, M., Korzun-Chłopicka, U., Szatkowski, M., Działo, M., Szopa, J. & Kulma, A. (2017). Optimization of phenolic compounds extraction from flax shives and their effect on human fibroblasts. *Evid Based Complement Alternat Med*, ID3526392.

Dalonso, N. & Petkowicz, C. L. (2012). Guarana powder polysaccharides: characterisation and evaluation of the antioxidant activity of a pectic fraction. *Food Chem, 134* (IV), 1804-1812.

Dihl, R. R. & Lehmann, F. K. M. (2011). *Skin study*. 1 ed. Canoas: ULBRA.

Engers, V. K., Behling, C. S. & Frizzo, M. N. (2011). The influence of oxidative stress in cellular aging process. *Rev Contexto Saúde, 11* (XX), 93-102.

Fabi, S. & Sundaram, H. (2014). The potential of topical and injectable growth factors and cytokines for skin rejuvenation. *Facial Plast Surg, 30* (II), 157-171.

Fukuoka, C. Y., Simões, A., Uchiyama, T., Arana-Chavez, V. E., Abiko, Y., Kuboyama, N. & Bhawal, U. K. (2017). The effects of low-power laser irradiation on inflammation and apoptosis in submandibular glands of diabetes-induced rats. *Plos One, 12* (I), e0169443.

Furukawa, S., Fujita, T., Shimabukuro, M., Iwaki, M., Yamada, Y., Nakajima, Y., Nakayama, O., Makishima, M., Matsuda, M. & Shimomura, I. (2004). Increased oxidative stress in obesity and its impact on metabolic syndrome. *J Clin Invest, 114* (XII), 1752-1761.

Ginani, F., Soares, D. M., Rocha, H. A. O., de Souza, L. B. & Barboza, C. A. G. (2018). Low-level laser irradiation induces *in vitro* proliferation of stem cells from human exfoliated deciduous teeth. *Lasers Med Sci, 33* (I), 95-102.

Guerreiro, R., Santos-Costa, Q. & Azevedo-Pereira, J. M. (2011). The chemokines and their receptors: characteristics and physiological functions. *Acta Med Port, 24*, 967-976.

Han, H., Roan, F. & Ziegler, S. F. (2017). The atopic march: current insights into skin barrier dysfunction and epithelial cell-derived cytokines. *Immunol Rev, 278* (I), 116-130.

Harris, M. I. N. C. (2009). Skin: structure, properties and aging. 3 ed. Sao Paulo: Senac.

Hertz, E., Cadoná, F. C., Machado, A. K., Azzolin, V., Holmrich, S., Assmann, C., Ledur, P., Ribeiro, E. E., de Souza Filho, O. C., Cattani, M. F. M. & da Cruz, I. B. M. (2015). Effect of Paullinia cupana on MCF-7 breast cancer cell response to chemotherapeutic drugs. *Mol Clin Oncol, 3* (I), 37-43.

Höhn, A., Weber, D., Jung, T., Ott, C., Hugo, M., Kochlik, B., Kehm, R., König, J., Grune, T. & Castro, J. P. (2017). Happily (n)ever after: Aging in the context of oxidative stress, proteostasis loss and cellular senescence. *Redox Biol, 11*, 482-501.

Houreld, N. & Abrahamse, H. (2007). Irradiation with a 632.8 nm helium-neon laser with 5 J/cm^2 stimulates proliferation and expression of interleukin-6 in diabetic wounded fibroblast cells. *Diabetes Technol Ther, 9* (V), 451-459.

Kamizato, K. K. & Brito, S. G. (2014). *Facial aesthetic techniques.* 1 ed. Sao Paulo: Érica.

Kober, H., Tatsch, E., Torbitz, V. D., Cargnin, L. P., Sangoi, M. B., Bochi, G. V., da Silva, A. R., Barbisan, F., Ribeiro, E. E., da Cruz, I. B. & Moresco, R. N. (2016). Genoprotective and hepatoprotective effects of guarana (Paullinia cupana Mart. var. sorbilis) on CCl4-induced liver damage in rats. *Drug Chem Toxicol, 39* (I), 48-52.

Krewer, C. C., Ribeiro, E. E., Ribeiro, E. A., Moresco, R. N., da Rocha, M. I., Montagner, G. F., Machado, M. M., Viegas, K., Brito, E. & da Cruz, I. B. (2011). Habitual intake of guaraná and metabolic morbidities: an epidemiological study of an elderly amazonian population. *Phytother Res, 25* (IX), 1367-1374.

Krewer, C. C., Suleiman, L., Duarte, M. M. M. F., Ribeiro, E. E., Mostardeiro, C. P., Montano, M. A. E., da Rocha, M. I. U. M., Algarve, T. D., Bresciani, G. & da Cruz, I. B. M. (2014). Guaraná, a supplement rich in caffeine and catechin, modulates cytokines: evidence from human *in vitro* and *in vivo* protocols. *Eur Food Res Technol, 239* (I), 49-57.

Kurutas, E. B. (2016). The importance of antioxidants which play the role in cellular response against oxidative/nitrosative stress: current state. *Nutr J, 15* (I), 71.

Lago, J. C. (2016). *Gene expression profile analysis from biomarkers of skin aging process.* Thesis (Doctorate in Medical Sciences) - Campinas State University, Campinas, SP, Brazil.

Lins, R. D. A. U., Dantas, E. M., Lucena, K. C. R., Catão, M. H. C. V., Granville-Garcia, A. F. & Neto, L. G. C. (2010). Biostimulation effects of low-power laser in the repair process. *Braz Annals Dermatol, 85* (VI), 849-855.

Luca, C., Pires, M. C. C. L., Corazza, S. & Higuchi, C. T. (2013). The role of genetics on cosmetology treatments antiaging. *InterfacEHS*, *8* (II), 63-91.

Machado, A. K. (2014). *Cito-genomic effect of hydrogen peroxide and guaraná (Paullinia cupana) in mesenchymal stem cells.* Dissertation (Master's degree in Pharmacology) - Federal University of Santa Maria, Santa Maria, RS, Brazil.

Machado, A. K., Cadoná, F. C., Azzolin, V. F., Dornelles, E. B., Barbisan, F., Ribeiro, E. E., Cattani, M. F. M., Duarte, M. M. M. F., Saldanha, J. R. P. & da Cruz, I. B. M. (2015). Guaraná (Paullinia cupana) improves the proliferation and oxidative metabolism of senescent adipocyte stem cells derived from human lipoaspirates. *Food Res Intern*, *67*, 426-433.

Maldaner, D. R., Pellenz, N. L., Barbisan, F., Azzolin, V. F., Mastella, M. H., Teixeira, C. F., Duarte, T., Ribeiro, E. A. M., Da cruz, I. B. M. & Duarte, M. M. F. (2019). Interaction between low-level laser therapy and Guarana (*Paullinia cupana)* extract potentializes antioxidant, anti-inflammatory, antiapoptotic and proliferative effects on dermal fibroblasts. *J Cosmet Dermatol-US*, [No Prelo].

McMorrow, J. P. & Murphy, E. P. (2011). Inflammation: a role for NR4A orphan nuclear receptors. *Biochem Soc Trans*, *39* (II), 688-693.

Montagner, G. F. F. S. (2010). *In vitro effect of ALA16VAL polimorphism of superoxide dismutase enzyme manganese dependent in oxidative metabolism of limphocytes.* Dissertation (Master's degree in Toxicological Biochemistry) - Federal University of Santa Maria, Santa Maria, RS, Brazil.

Nestor, M., Andriessen, A., Berman, B., Katz, B. E., Gilbert, D., Goldberg, D. J., Gold, M. H., Kirsner, R. S. & Lorenc, P. Z. (2017). Photobiomodulation with non-thermal lasers: mechanisms of action and therapeutic uses in dermatology and aesthetic medicine. *J Cosmet Laser Ther*, *19* (IV), 190-198.

Neves, R. P. P., Fernandes, P. A., Varandas, A. J. C. & Ramos, M. J. (2014). Benchmarking of density functionals for the accurate description of thiol-disulfide exchange. *J Chem Theory Comput*, *10* (XI), 4842-4856.

Otobone, F. J., Sanches, A. C., Nagae, R., Martins, J. V., Sela, V. R., de Mello, J. C. & Audi, E. A. (2007). Effect of lyophilized extracts from guaraná seeds [Paullinia cupana var. sorbilis (Mart.) Ducke] on behavioral profiles in rats. *Phytother Res, 21* (VI), 531-535.

Peixoto, H., Roxo, M., Röhrig, T., Richling, E., Wang, X. & Wink, M. (2017). Anti-aging and antioxidant potential of Paullinia cupana var. sorbilis Mart var. sorbilis: findings in Caenorhabditis elegans indicate a new utilization for roasted seeds of guarana. *Medicines, 4* (III).

Pereira, M. F. L. (2013). *Technical resources in aesthetics II.* Sao Paulo: Difusão.

Perl, A. (2013). Oxidative stress in the pathology and treatment of systemic lupus erythematosus. *Nat Rev Rheumatol, 9* (XI), 674-686.

Pinto, M. V. M. (2011). *Phototherapy: clinical aspects of rehabilitation.* 1 ed. Sao Paulo: Andreoli.

Piva, J. A. A. C., Abreu, E. M. C., Silva, V. S. & Nicolau, R. A. (2011). Effect of low-level laser therapy on the initial stages of tissue repair: basic principles. *Braz Annals Dermatol, 86* (V), 947-954.

Poljšak, B. & Dahmane, R. (2012). Free radicals and extrinsic skin aging. *Dermatol Res Pract*, ID135206.

Portella, R. B. (2011). Evaluation of the polymorphism in the interleukin-10 gene in peri-implant disease. Dissertation (Master's degree in Odontology) - University Veiga de Almeida, Rio de Janeiro, RJ, Brazil.

Portella, R. L., Barcelos, R. P., da Rosa, E. J. F., Ribeiro, E. E., da Cruz, I. B. M., Suleiman, L. & Soares, F. A. A. (2013). Guaraná (Paullinia cupana Kunth) effects on LDL oxidation in elderly people: an *in vitro* and *in vivo* study. *Lipids Health Dis, 12*.

Quan, T. & Fisher, G. J. (2015). Role of age-associated alterations of the dermal extracellular matrix microenvironment in human skin aging. *Gerontology, 61* (V), 427-434.

Rambo, C. S. M. (2013). *Analysis of low power laser therapy (660nm) on the expression of inflammatory biomarkers in the healing process: a comparative study between young and elderly rats.* Dissertation (Master's degree in Rehabilitation Sciences) - University Nine of July, Sao Paulo, SP, Brazil.

Ramos, F. S., Maifrino, L. B. M., Alves, S., Alves, B. C. A., Perez, M. M., Feder, D., Azzalis, L. A., Junqueira, V. B. C. & Fonseca, F. L. A. (2018). The effects of transcutaneous low-level laser therapy on the skin healing process: an experimental model. *Lasers Med Sci, 33* (V), 967-976.

Rapisarda, V., Borghesan, M., Miguela, V., Encheva, V., Snijders, A. P., Lujambio, A. & O'Loghlen, A. (2017). Integrin beta 3 regulates cellular senescence by activating the TGF-β pathway. *Cell Rep, 18* (X), 2480-2493.

Sies, H., Berndt, C. & Jones, D. P. (2017). Oxidative stress. *Annu Rev Biochem, 86,* 715-748.

Silva, V. S., Abreu, E. M. C., Nicolau, R. A. & Soares, C. P. (2016). Comparative analysis of different doses of coherent light (laser) and non-coherent light (light-emitting diode) on cellular necrosis and apoptosis: a study *in vitro. Res Biomed Eng, 32* (IV), 372-379.

Silveira, P. C. L., Silva, L. A., Tuon, T., Freitas, T. P., Streck, E. L. & Pinho, R. A. (2009). Effects of low-level laser therapy on epidermal oxidative response induced by wound healing. *Braz J Phys Ther, 13* (IV), 281-287.

Sinha, K., Das, J., Pal, P. B. & Sil, P. C. (2013). Oxidative stress: the mitochondria-dependent and mitochondria-independent pathways of apoptosis. *Arch Toxicol, 87* (VII), 1157-1180.

Souza, J. R. M., Oliveira, R. T., Blotta, M. H. S. L. & Coelho, O. R. (2008). Serum levels of interleukin-6 (IL-6), interleukin-18 (IL-18) and C-reactive protein (CRP) in patients with type-2 diabetes and acute coronary syndrome without ST-segment elevation. *Brazilian Archives of Cardiology, 90* (II), 86-90.

Souza, R. O. (2015). *In vitro and in vivo evaluation of photochemoprotective activity of Byrsonima crassifolia fraction and (+) catechin against the damages induced by UVB radiation.* Thesis (Doctorate in Pharmaceutical Sciences) - University of Sao Paulo, Ribeirao Preto, SP, Brazil.

Suleiman, L., Barbisan, F., Ribeiro, E. E., Moresco, R. N., Bochi, G., Duarte, M. M. M. F., Antunes, K. T., Cattani, M. F. M., Unfer, T. C., Azzolin, V. F., Griner, J. & da Cruz, I. B. M. (2016). Guaraná supplementation modulates triglycerides and some metabolic blood biomarkers in overweight subjects. *Ann Obes Disord, 1* (I).

Szezerbaty, S. K. F., de Oliveira, R. F., Pires-Oliveira, D. A. A., Soares, C. P., Sartori, D. & Poli-Frederico, R. C. (2018). The effect of low-level laser therapy (660 nm) on the gene expression involved in tissue repair. *Lasers Med Sci, 33* (II), 315-321.

Szwed, D. N. & Santos, V. L. P. (2016). Growth factors involved in skin healing. *Cad Esc de Sau, 1* (XV), 7-17.

Teixeira, I. N. D. O. & Guariento, M. E. (2010). Biology of aging: theories, mechanisms, and perspectives. *J Science and Collective Health, 15* (VI), 2845-2857.

Thomson, J., Singh, M., Eckersley, A., Cain, S. A., Sherratt, M. J. & Baldock, C. (2019). Fibrillin microfibrils and elastic fibre proteins: Functional interactions and extracellular regulation of growth factors. *Semin Cell Dev Biol, 89*, 109-117.

Velarde, M. C. (2017). Epidermal barrier protects against age-associated systemic inflammation. *J Invest Dermatol, 137* (VI), 1206-1208.

Veskoukis, A. S., Tsatsakis, A. M. & Kouretas, D. (2012). Dietary oxidative stress and antioxidant defense with an emphasis on plant extract administration. *Cell Stress Chap, 17* (I), 11-21.

Vitale, G., Salvioli, S. & Franceschi, C. (2013). Oxidative stress and the ageing endocrine system. *Nat Rev Endocrinol, 9* (IV), 228-240.

Xu, X., Zheng, L., Yuan, Q., Zhen, G., Crane, J. L., Zhou, X. & Cao, X. (2018). Transforming growth factor-β in stem cells and tissue homeostasis. *Bone Res, 6*, 2.

Yamaguti-Sasaki, E., Ito, L. A., Canteli, V. C., Ushirobira, T. M., Ueda-Nakamura, T., Dias Filho, B. P., Nakamura, C. V. & de Mello, J. C. (2007). Antioxidant capacity and *in vitro* prevention of dental plaque formation by extracts and condensed tannins of Paullinia cupana. *Molecules, 12* (VIII), 1950-1963.

Yang, L., Zhang, D., Wu, H., Xie, S., Zhang, M., Zhang, B. & Tang, S. (2018). Basic fibroblast growth factor influences epidermal homeostasis of living skin equivalents through affecting fibroblast phenotypes and functions. *Skin Pharmacol Physiol, 31* (V), 229-237.

Yu, L. S., Li, X. B., Fan, Y. Y., Ye, G. H., Li, J. L., Fu, T. T., Liao, Y. & Li, F. (2017). Expression of KGF-1 and KGF-2 in skin wounds and its application in forensic pathology. *Am J Forensic Med Pathol, 38* (III), 199-210.

Yun, S. H. & Kwok, S. J. J. (2017). Light in diagnosis, therapy and surgery. *Nat Biomed Eng, 1*.

Zimmermann, M., Koreck, A., Meyer, N., Basinski, T., Meiler, F., Simone, B., Woehrl, S., Moritz, K., Eiwegger, T., Schmid-Grendelmeier, P., Kemeny, L. & Akdis, C. A. (2011). TNF-like weak inducer of apoptosis (TWEAK) and TNF-α cooperate in the induction of keratinocyte apoptosis. *J Allergy Clin Immunol, 127* (I), 200-207.

INDEX

A

absorption spectra, 18
acetone, 74, 76, 82
adaptive immune responses, 116
adenocarcinoma, 66, 67, 68
adenosine, viii, 1, 2, 113
adenosine triphosphate, viii, 1, 3, 113
adipose tissue, 25, 79
adult stem cells, 79
aesthetic(s), 72, 126, 129, 130, 131
angiogenesis, viii, 2, 9, 10, 11, 25, 29, 54, 57, 61, 63, 71, 78, 87, 97, 117
antioxidant, x, 27, 70, 75, 78, 79, 81, 82, 83, 88, 89, 92, 97, 111, 114, 118, 119, 121, 122, 123, 124, 125, 127, 130, 131, 133
antitumor, 72, 81, 122
apoptosis, 19, 26, 59, 62, 110, 111, 113, 115, 120, 127, 128, 132, 134

B

basement membrane, viii, 52, 54, 55
beneficial effect, 39, 97
benign, ix, 52, 56, 60, 61, 111
biochemistry, 37, 47, 49
biological activities, 81
biological processes, 12, 15, 119
biological responses, 3, 15
biological systems, 21, 84, 111, 113
biomarkers, 68, 120, 129, 131, 133
blood vessels, 10, 54, 94, 95, 96, 97
bone marrow, 12, 30, 59
bradykinin, 74, 119
breast cancer, 67, 128

C

Ca^{2+}, 4, 6, 19, 22, 37, 43
caffeine, 78, 121, 129
calcium, viii, 2, 3, 29, 42
cancer, vii, viii, ix, 23, 31, 39, 46, 51, 52, 53, 54, 55, 56, 57, 58, 62, 63, 64, 65, 66, 67, 68, 75, 125
cancer cells, 54, 57, 62
cancer progression, ix, 52, 62, 63
cancer-associated fibroblasts (CAFs), vii, ix, 52, 54, 55, 56, 57, 58, 59, 60, 62, 63, 65
cancer-related fatigue, 126
carcinogenesis, viii, 51, 55
carcinoma, viii, 52, 67

cardiovascular disease, 122
caspases, 123, 124
cell biology, 31, 37, 47
cell culture, 9, 79, 83, 84, 85, 120
cell cycle, 23, 49, 116, 123, 124
cell death, viii, 2, 3, 6, 29, 54, 110, 111, 114, 115
cell division, 110, 111
cell metabolism, 28, 119
cell organelles, 112
cell signaling, 123, 124
cell surface, 20, 111
cellular energy, 19, 62
chemokines, ix, 9, 10, 52, 59, 65, 128
chemotaxis, 26, 32, 59, 117
chemotherapeutic agent, 62
chemotherapy, ix, 52, 57, 62, 63, 67
chemotherapy treatment, ix, 52, 62
collagen, 8, 9, 11, 12, 21, 26, 27, 28, 36, 43, 71, 72, 75, 112, 116, 117, 120
connective tissue, 8, 11, 12, 43, 54, 60, 64, 112
cysteine, 60, 61, 62
cytochrome, 2, 17, 19, 37, 38, 40, 43, 48, 49, 118
cytokines, ix, 3, 8, 9, 10, 11, 31, 49, 52, 85, 93, 111, 112, 115, 116, 119, 123, 124, 128, 129
cytotoxicity, 75, 78, 125, 127

D

degradation, 12, 22, 26, 111, 114, 116, 122
dendritic cell, 24, 25, 59
dermatology, 40, 43, 130
dermis, 8, 110, 116
diabetes, 10, 23, 40, 120, 128, 129, 132
diode laser, 19, 28, 33, 36
DNA, 13, 18, 20, 25, 26, 79, 84, 85, 110, 111, 113, 115, 116, 119, 122, 123, 126
DNA damage, 13, 79, 110, 122

drug resistance, 57, 62, 65
drugs, 44, 59, 72, 128

E

earthworms, x, 70, 81, 83, 86, 87, 94, 96
ECM, 8, 10, 11, 12, 20, 26, 28, 39, 57, 60
embryonic stem cells, 79, 81
endometriosis, 55, 65
endothelial cells, ix, 9, 39, 52, 59, 60, 61, 116, 117, 118
energy, 7, 13, 15, 17, 19, 28, 113, 118, 119
energy density, 15, 17, 119
environmental factors, 112, 114
enzymatic activity, 114
enzyme, 4, 19, 58, 85, 92, 110, 118, 123, 130
enzyme-linked immunosorbent assay (ELISA), 85
enzymes, 11, 18, 27, 50, 59, 79, 85, 92, 110, 111, 113, 114, 118, 124
epithelial cells, 24, 45, 53, 54, 59, 71, 117
epithelial ovarian cancer, 64, 65, 68
epithelium, viii, 52, 53, 55, 60, 117
estrogen, 60, 66
ethyl alcohol, 83, 86
exposure, x, 17, 38, 46, 60, 63, 92, 110, 111, 114, 119, 123
external environment, ix, 70, 111
extracellular matrix, ix, 8, 14, 25, 30, 31, 33, 45, 46, 52, 54, 55, 57, 58, 59, 112, 117, 131
extraction, 76, 77, 82, 127
extracts, 72, 73, 74, 76, 104, 121, 131, 133

F

fibroblast growth factor, 33, 34, 45, 61, 85, 117, 133
fibroblast proliferation, viii, 2, 10, 14, 23, 24, 29, 119

Index

fibroblast(s), v, vii, viii, ix, x, 1, 2, 3, 4, 5, 6, 7, 8, 10, 11, 12, 14, 16, 18, 19, 20, 21, 22, 23, 24, 26, 27, 28, 29, 30, 31, 32, 33, 34, 35, 36, 37, 39, 40, 41, 42, 43, 44, 45, 46, 47, 48, 50, 51, 52, 54, 55, 56, 57, 58, 59, 60, 61, 62, 63, 64, 65, 66, 67, 68, 70, 71, 75, 78, 80, 82, 84, 85, 88, 89, 90, 92, 93, 104, 110, 111, 116, 117, 118, 121, 123, 127, 130
fibrogenesis, 26, 42, 46
fibrosis, 10, 27, 39, 49, 63, 66

G

gene expression, 3, 6, 19, 20, 22, 37, 48, 80, 110, 117, 123, 124, 125, 133
genes, 20, 21, 23, 24, 25, 48, 60, 79, 81
genetic factors, x, 110
gingival, 20, 23, 28, 32, 33, 36, 41, 45
glucose, 25, 34, 62, 76
glutathione, 27, 62, 85, 92, 114
growth factor, ix, 7, 8, 9, 10, 11, 12, 15, 25, 26, 28, 36, 39, 43, 45, 47, 48, 49, 52, 57, 59, 60, 61, 64, 65, 75, 85, 89, 117, 123, 124, 128, 133

H

healing, vii, ix, 1, 8, 9, 10, 12, 23, 24, 26, 31, 34, 41, 42, 44, 47, 48, 63, 70, 72, 75, 76, 77, 78, 79, 80, 81, 89, 93, 97, 117, 119, 120, 131, 132, 133
herbal medicine, 77, 80
homeostasis, 8, 9, 19, 21, 23, 25, 112, 114, 115, 133
human lung fibroblasts, 40
human skin, 19, 27, 28, 34, 36, 37, 43, 47, 50, 115, 131
hydrogen, 14, 19, 84, 113, 130
hydrogen peroxide, 14, 84, 113, 130
hypoxia, 10, 19, 60

I

IFN, 47, 85, 93, 115, 116, 126
immune response, 63, 110, 116
in vitro, x, 3, 9, 16, 18, 19, 20, 25, 26, 27, 28, 31, 39, 42, 46, 47, 49, 50, 60, 70, 76, 79, 80, 82, 83, 84, 86, 88, 90, 97, 120, 122, 124, 128, 129, 131, 132, 133
in vitro environment, 9
in vivo, x, 3, 9, 16, 22, 26, 47, 70, 80, 81, 82, 83, 88, 113, 122, 129, 131, 132
inflammation, vii, ix, 1, 9, 10, 14, 29, 33, 49, 52, 55, 58, 71, 73, 81, 111, 115, 116, 128, 133
inflammatory cells, ix, 14, 52, 57, 58
inflammatory disease, 25
inflammatory mediators, 9, 10
infrared spectroscopy, 32
inhibition, 11, 23, 71, 75, 117
inhibitor, 8, 22, 41
irradiation, vii, 1, 2, 4, 5, 6, 7, 13, 14, 16, 19, 22, 27, 28, 30, 32, 36, 37, 39, 41, 42, 45, 46, 47, 48, 49, 50, 119, 120, 123, 128
ischemia, 33, 72

J

jaundice, 10

K

keratinocyte(s), 9, 10, 11, 12, 41, 44, 47, 67, 71, 80, 85, 111, 117, 126, 134

L

Langerhans cells, 59
lasers, 2, 13, 27, 30, 32, 33, 36, 37, 39, 41, 42, 44, 45, 46, 48, 118, 127, 128, 130, 132, 133

LDL, 122, 131
lesions, 54, 56, 75, 126, 127
leukocytes, 10, 116
ligand, 24, 30, 59
light emitting diode(LED), vii, 1, 2, 6, 14, 32, 127
lipid peroxidation, 113, 122
liposuction, 79, 122
liquid chromatography, 121
lymphocytes, 10, 25, 42

M

macrophages, 9, 10, 25, 57, 59, 115, 117, 118, 119
mast cells, 117, 119
matrix metalloproteinase, 31
mesenchymal stem cells, 45, 130
messenger ribonucleic acid (mRNA, 7, 23, 28, 36, 47, 61, 116, 120, 126
meta-analysis, 16, 126
metabolism, vii, 23, 25, 49, 54, 62, 93, 112, 113, 114, 122, 123, 130
metastasis, 54, 57, 58, 60, 61
migration, x, 7, 9, 10, 12, 14, 21, 24, 28, 35, 41, 44, 60, 70, 71, 80, 84, 85, 89, 90, 114, 117, 120
mitochondria, 3, 8, 14, 16, 18, 19, 22, 50, 112, 113, 126, 132
mitogen, 21, 28, 116
MMPs, viii, 2, 4, 6, 19, 21, 22, 29, 33, 111, 116
models, x, 70, 72, 74, 75, 78, 80, 81, 82, 97, 120
molecular medicine, 44, 46
molecules, 11, 13, 18, 54, 58, 72, 74, 76, 78, 82, 110, 114
monoclonal antibody, 60
myofibroblasts, 8, 11, 22, 26, 27, 28, 37, 43, 71

N

Na^+, 22
NADH, 17, 19, 118
National Academy of Sciences, 34, 67
necrosis, 19, 72, 132
neoplastic tissue, 61, 120
neurons, 49, 59
neuropeptides, 20
neuroprotection, 38
neutrophils, 9, 59, 115
NIR, 3, 12, 13, 18
nitric oxide, viii, 1, 2, 11, 19, 43, 49, 114
non-cancerous cells, 62
nuclear receptors, 47, 130

O

obesity, 25, 34, 125, 128
organs, xi, 11, 58, 79, 81, 110
ovarian cancer, v, vii, ix, 51, 52, 53, 55, 56, 57, 60, 62, 63, 64, 65, 66, 67, 68
ovarian tumor, viii, 51, 56, 57, 58, 60, 61
oxidation, 14, 18, 38, 114, 122, 123, 131
oxidative damage, 113, 114
oxidative stress, 27, 30, 33, 40, 46, 79, 85, 111, 112, 114, 115, 116, 122, 125, 126, 127, 128, 129, 133
oxygen, 15, 18, 19, 113, 115

P

P13K/AKT, 23
p53, 116, 120
paclitaxel, 62, 63
pancreatic cancer, 60
paralysis, 86
parenchyma, 57
Parkinsonism, 38
pathogenesis, 55, 65, 66, 112

Index

pathology, 37, 41, 44, 46, 64, 66, 119, 131, 134
pathophysiology, 125
peripheral blood mononuclear cell, 125
photobiomodulation, v, vii, 1, 2, 12, 18, 30, 34, 35, 36, 38, 39, 40, 43, 48, 49, 50, 118, 127, 130
PI3K, 25, 28, 35, 42, 45, 65
PI3K/AKT, 28
pro-inflammatory, 3, 22, 26, 85, 93, 115
proliferation, viii, x, 2, 3, 4, 5, 6, 7, 8, 9, 10, 11, 12, 18, 23, 24, 26, 27, 28, 29, 36, 40, 42, 43, 49, 54, 57, 59, 63, 65, 70, 71, 79, 80, 85, 89, 114, 117, 120, 122, 123, 124, 128, 129, 130
proteins, 12, 18, 20, 23, 25, 54, 110, 112, 113, 114, 116, 133
proteolytic enzyme, 12
proto-oncogene, 23
pulp, 34, 48, 121

R

radiation, 12, 13, 38, 41, 47, 111, 118, 126, 132
radicals, 14, 113, 131
reactive oxygen, 3, 32, 35, 79, 111
receptor(s), 15, 20, 24, 25, 26, 30, 32, 44, 47, 49, 59, 61, 65, 111, 128
regeneration, x, 10, 12, 14, 18, 31, 33, 36, 70, 71, 79, 81, 82, 87, 88, 93, 94, 96, 97, 110, 120
repair, 2, 8, 11, 12, 31, 33, 34, 36, 49, 66, 71, 117, 119, 120, 129, 131, 133
resistance, 57, 59, 62, 75, 122
respiration, 5, 19, 114, 118
response, vii, 1, 5, 14, 15, 16, 19, 20, 25, 30, 34, 37, 55, 57, 59, 61, 62, 63, 115, 116, 119, 128, 129, 132

S

senescence, xi, 8, 24, 79, 110, 111, 122, 125, 126, 129, 132
signaling pathway, viii, 2, 3, 15, 20, 21, 23, 24, 29, 46, 49, 67, 74
signalling, 30, 41, 42, 65, 67
signals, 11, 19, 20, 54, 114, 117
skin, vii, ix, 8, 9, 12, 14, 19, 23, 25, 27, 30, 34, 41, 48, 49, 70, 72, 75, 78, 81, 111, 112, 114, 115, 116, 118, 120, 126, 127, 128, 129, 131, 132, 133, 134
skin cancer, 114
stimulation, 2, 15, 23, 111, 116, 117, 119, 120
stress, 27, 45, 46, 71, 79, 115, 125, 129, 131, 132, 133
stress response, 45
stressors, xi, 110, 111
stroma, ix, 52, 54, 55, 56, 57, 58, 59, 60, 61, 63, 64, 65, 66, 67, 68
stromal cells, 55, 56, 61, 65
structural changes, 14, 112
structure, 3, 8, 9, 11, 15, 31, 113, 128
subcutaneous tissue, 112
surgical removal, x, 70, 82, 86, 87, 93, 94, 96
synthesis, 8, 9, 11, 14, 18, 21, 23, 26, 27, 28, 39, 72, 75, 85, 117, 119, 120
systemic lupus erythematosus, 126, 131

T

T cells, 49, 59, 67
T lymphocytes, 59, 119
terpenes, 82, 121
TGF, 6, 7, 8, 20, 26, 28, 30, 35, 39, 43, 45, 46, 49, 59, 60, 64, 75, 111, 112, 117, 132
therapeutic agents, 72
therapeutic approaches, vii, viii, 2, 3
therapeutic targets, ix, 52, 60

therapy, 2, 14, 29, 30, 32, 33, 34, 36, 37, 38, 46, 47, 60, 62, 67, 68, 119, 121, 123, 127, 130, 131, 132, 133, 134
TIMP, 8, 33
TIMP-1, 33
TIMP-2, 33
TNF-α, 74, 85, 93, 115, 116, 119, 123, 124, 126, 134
toxicity, 13, 27, 34, 122, 125
transcription factors, viii, 2, 3, 16, 23, 25, 29
transforming growth factor, 8, 36, 40, 45, 46, 47, 48, 59, 111, 117
transition metal ions, 113
traumatic brain injury, 36
treatment, ix, 2, 13, 14, 15, 31, 33, 39, 40, 44, 52, 54, 58, 59, 62, 75, 78, 85, 86, 87, 88, 89, 91, 93, 120, 122, 123, 124, 126, 131
tumor(s), ix, 52, 53, 54, 55, 56, 57, 58, 59, 60, 61, 63, 64, 65, 66, 67, 68, 74, 115, 116
tumor cells, 56, 58, 59
tumor development, 55, 57
tumor growth, 60, 61, 63
tumor microenvironment, ix, 52, 54, 57, 63, 65, 68
tumor progression, 55, 59, 61
tumor stroma, 56, 57, 58, 60
type 1 collagen, 48
type 2 diabetes, 25

V

vascular endothelial growth factor (VEGF), 8, 117
vascularization, 61, 87, 94, 112
vasoconstriction, 115
vasodilation, 115, 116
VEGF, 28, 39, 120
VEGF expression, 120
vitamin D, 14

W

wound healing, vii, viii, ix, x, 2, 3, 6, 9, 10, 11, 12, 13, 14, 23, 24, 26, 29, 30, 31, 32, 33, 34, 35, 36, 37, 39, 40, 42, 43, 44, 46, 48, 49, 52, 55, 70, 71, 72, 77, 78, 80, 81, 82, 90, 104, 127, 132

Y

yttrium, 27

Z

zinc, 18